Chemical Searching
on an Array Processor

COMPUTERS AND CHEMICAL STRUCTURE INFORMATION SERIES

Series Editor: **Dr. D. Bawden,** *Department of Information Science, The City University, London, England*

1. Three-Dimensional Chemical Structure Handling
   Peter Willett

2. Chemical Searching on an Array Processor
   Terence Wilson

3. Three-Dimensional Chemical Similarity Searching
   Catherine Pepperrell

Series of related interest:

CHEMOMETRICS SERIES

9. Molecular Connectivity in Structure-Activity Analysis
   Lemont B. Kier *and* **Lowell H. Hall**

11. Chemical Pattern Recognition
    O. Štrouf

12. Similarity and Clustering in Chemical Information Systems
    Peter Willett

13. Multivariate Chemometrics in QSAR: A Dialogue
    Peter P. Mager

14. Application of Pattern Recognition to Catalytic Research
    I. I. Ioffe

15. Distance Geometry and Molecular Conformation
    G. M. Crippen *and* **T. F. Havel**

16. Design Statistics in Pharmacochemistry
    Peter P. Mager

Chemical Searching on an Array Processor

Terence Wilson
Molecular Design Ltd., San Leandro, USA

RESEARCH STUDIES PRESS LTD.
Taunton, Somerset, England

JOHN WILEY & SONS INC.
New York · Chichester · Toronto · Brisbane · Singapore

RESEARCH STUDIES PRESS LTD.
24 Belvedere Road, Taunton, Somerset, England TA1 1HD

Marketing and Distribution:

Australia and New Zealand:
Jacaranda Wiley Ltd.
GPO Box 859, Brisbane, Queensland 4001, Australia

Canada:
JOHN WILEY & SONS CANADA LIMITED
22 Worcester Road, Rexdale, Ontario, Canada

Europe, Africa, Middle East and Japan:
JOHN WILEY & SONS LIMITED
Baffins Lane, Chichester, West Susses, England

North and South America:
JOHN WILEY & SONS INC.
605 Third Avenue, New York, NY 10158, USA

South East Asia:
JOHN WILEY & SONS (SEA) PTE LTD.
37 Jalan Pemimpin 05-04
Block B Union Industrial Building, Singapore 2057

Library of Congress Cataloging-in-Publication Data

Wilson, Terence, 1967–
   Chemical searching on an array processor / Terence Wilson.
     p.   cm. — (Computers and chemical structure information
series; 2)
    Includes bibliographical references and index.
    ISBN 0-86380-128-5 (Research Studies Press Ltd.). —ISBN
0-471-93436-4 (John Wiley & Sons, Inc.)
    1. Chemical structure—Data bases. 2. Information storage and
retrieval systems—Chemical structure. 3. Array processors.
I. Title. II. Series.
QD471.W63 1992
025.06′54′0285435—dc20   92-19978
                               CIP

British Library Cataloguing in Publication Data

A catalogue record for this book
is available from the British Library.

ISBN 0 86380 128 5 (Research Studies Press Ltd.)
ISBN 0 471 93436 4 (John Wiley & Sons Inc.)

Printed in Great Britain by Galliard (Printers) Ltd., Great Yarmouth

Preface

This book reports work carried out in the Department of Information Studies in the University of Sheffield into the processing of chemical structures by a parallel process known as the DAP. The PhD project was sponsored by the makers of the DAP, Active Memory Technology Ltd. of Reading and was supervised by Prof. Peter Willett.

The book describes a series of investigations centring on applications of the DAP to problems of searching representations of chemical structures. It begins by reviewing existing knowledge on chemical structure processing, on the underlying graph theory on which representations are based, the nature of algorithms, and the hardware and software of parallel computing devices. These reviews lay the basis for the description of extensive studies on the development and testing of a variety of algorithms which seek to find the best match between algorithms and device characteristics.

Acknowledgements

I would like to thank Professor Peter Willett for his enthusiastic support of the research described in this book. Thanks are also due to my industrial supervisor, Dr Stewart Reddaway, and his colleague Dr Kermen Mok, of AMT (Active Memory Technology Ltd.), for their help with the work during my visits to AMT in Reading. While in Sheffield, Dr John Steele and the staff of the Centre for Parallel Computing at Queen Mary and Westfield College provided invaluable support in using the DAP.

Thanks are due to Dr Geoff Downs for generating the connection tables that were used in the experiments, and to Barnard Chemical Information Ltd. for supplying the software used to draw the structure diagrams in this book. I would also like to express my gratitude to Dr Helen Grindley for providing code and data for the macromolecule ranking algorithm.

I would like to thank Dr John Holliday for taking the time to read the first draft of this book. Recognition should also go to Dr David Bawden, the editor, and Veronica Wallace of Research Studies Press, for their assistance in the publishing of the research reported in the book.

Finally, I owe most to my parents for their support and encouragement during my research*.

---

* This research was a Science and Engineering Council and co-operative project supported by Active Memory Technology Ltd.

vii

Introduction by the
Series Editor

One of the most fruitful aspects of research and development in science and technology is seen when newly developed informatics technologies are applied to an area where such advances are clearly needed. One such example is the application of supercomputers to molecular dynamics. Another is the use of advanced graphics displays for the visualisation of molecular, and biomolecular, structure. This book describes another such example: the application of parallel processing hardware to retrieval from databases of chemical structure information.

It is no mere coincidence that the examples quoted, including the present one, relate to the chemical domain. The application of informatics to chemistry has been particularly fruitful, and builds upon the well-developed information systems which have served chemistry, and related sciences, for more than a century. The readiness with which chemical knowledge may be systematised and formalised, by means of nomenclatures, notations and formulae, has provided an opportunity, unique among the sciences, for the application of systems and techniques of documentation and informatics. In particular, it is the ubiquitousness of the chemical structure diagram representation, a complete and unambiguous description of a substance, which may be both represented by computer graphics, and stored and retrieved according to the formalisms of graph theory, which underlies the sophisticated computer-based chemical information systems developed over the past decades. It is this unifying factor which forms the common link between the books of this series.

The handling of chemical information, on this basis, has provided a test-bed for many experimental and newly developed informatics systems, both hardware and software. The chemical information environment provides both a test for the new systems themselves, in an environment often somewhat removed from their designers' original conceptions, and also the potential of novel and valuable capabilities for chemists and chemical information specialists. In particular, advances made in chemical information systems have often been transferred at a later stage to more general information processing situations.

This book describes the application of a novel type of computer hardware, a Distributed Array Processor, providing a parallel processing capability, to chemical structure information. Parallel processing hardware, universally hailed as one of the fundamental components of future information processing systems, has made slower headway than expected in the database retrieval context. Terence Wilson's studies, reported here, give a unique insight, not merely into the theoretical and practical issues involved in the use of this kind of computer system for information retrieval. The value of this book goes beyond its exposition of the advantages, and associated development problems, of parallel hardware of this type in chemical information systems, valuable though these detailed and thorough analyses and experiments are. They also throw a more general light on the applicability of these systems for database processing in the wider context.

David Bawden
City University
London, May 1992

Contents

1 Introduction   **1**

  1.1 Objectives 1

  1.2 Outline 3

2 The Storage of Chemical Information   **5**

  2.1 The Evolution of Chemical Information Systems 5

  2.2 Modern Day Use of Chemical Information Systems 9

  2.3 Representation of Compounds 10

    2.3.1 The 2-D Structure Diagram 10

    2.3.2 Ambiguity in Representation 10

    2.3.3 Unambiguous Representation of Compounds 12

  2.4 Retrieval in Structure-Based Chemical Information Systems 16

    2.4.1 Structure Search 16

    2.4.2 Substructure Search 17

    2.4.3 Similarity Search 17

    2.4.4 Generic Structures and Search of Generic Structure Files . . 19

  2.5 Structure and Substructure Search 21

    2.5.1 Structure Search 21

    2.5.2 Substructure Search 23

  2.6 Current Chemical Information Systems 31

    2.6.1 Introduction 31

    2.6.2 The CAS ONLINE Search System 31

    2.6.3 MACCS Molecular Access Program 33

    2.6.4 The New Generation of Chemical Information Systems 34

3 Graph Theory and Complexity   **35**

   3.1   Graph Theory 35

         3.1.1   Introduction 35

         3.1.2   Formal Definitions 36

         3.1.3   Graph Theory and the Connection Table 36

         3.1.4   Structure/Substructure Search and Graph Theory 39

   3.2   Complexity 40

         3.2.1   Introduction 40

         3.2.2   Complexity Classes 42

         3.2.3   NP Completeness 43

         3.2.4   Hard Problems in Graph Theory 43

         3.2.5   Dealing With NP-Complete Problems 43

   3.3   Conclusion 45

4 Parallel Processing   **46**

   4.1   Models of Computation 46

         4.1.1   The Von Neumann Model of Computation 46

         4.1.2   Disadvantages of the Von Neumann Model 47

         4.1.3   A More General Model of Computation 49

   4.2   Achieving Parallelism in Hardware 51

         4.2.1   Pipelining 51

         4.2.2   Spatial Replication 52

   4.3   Other Taxonomies 64

         4.3.1   Shore's Taxonomy 64

         4.3.2   ASN (Algebraic Structure Notation) 65

   4.4   Hardware Performance Measurements 65

   4.5   Parallel Algorithms 66

         4.5.1   Concurrency Control 67

         4.5.2   Granularity 67

         4.5.3   Communication Geometry 67

         4.5.4   Measuring Algorithm Performance 69

         4.5.5   Amdahl's Law 71

5 DAP Hardware and Software   **73**

   5.1 Introduction 73

   5.2 Dap Architecture 74

       5.2.1 The Master Control Unit 74

       5.2.2 The Processor Array 75

       5.2.3 DAP I/O. 80

   5.3 Program Development on the DAP 81

       5.3.1 Fortran Plus 82

       5.3.2 The Structure of a DAP Program 89

       5.3.3 The Parallel Data Transform Library 91

6 The Ullmann Algorithm   **93**

   6.1 Background 93

   6.2 Finding Isomorphisms 93

   6.3 The Ullmann Heuristic 94

       6.3.1 Terminology 94

   6.4 Implementation for Chemical Structures 97

       6.4.1 The Search Stage 98

       6.4.2 Calculating the M Matrix 99

   6.5 A Parallel Ullmann Algorithm 100

       6.5.1 Approaches to Parallelism 100

       6.5.2 An Algorithmic Technique 101

       6.5.3 Complexity 104

       6.5.4 Test Data 105

       6.5.5 Initial Experiments 105

       6.5.6 Conclusions 106

   6.6 Data-Parallel Ullmann 106

       6.6.1 Introduction 106

       6.6.2 Data Mapping 107

       6.6.3 Algorithm Design 107

       6.6.4 Relaxation Procedure 111

       6.6.5 Tree Compression 112

       6.6.6 Algorithm Inefficiencies 114

   6.7 Main Experiments (Algorithms I & II) 116

6.7.1   Database Structures 116

6.7.2   Query Structures 117

6.7.3   Results 117

6.8   Hybrid Algorithm 119

6.9   Algorithm IV: The Drip-Feed Algorithm 120

6.10 Conclusions 121

7   A DAP-Based Substructure-Search System   127

7.1   Introduction 127

7.2   Background 127

7.3   System Design 128

7.4   Structure Input 128

7.5   Design of a Parallel Search System 130

     7.5.1   Data Storage 130

7.6   Generating Binary Structure Files 133

7.7   Compression Algorithm 134

     7.7.1   Data Input 134

     7.7.2   Compression Stage 135

     7.7.3   Bit-String Concatenation 136

     7.7.4   Bit-String Record Size 137

7.8   Run-Time System 138

     7.8.1   Data Mapping 138

     7.8.2   Bond Matrix Generation 138

     7.8.3   Generating Tentative Mappings 140

     7.8.4   System Performance 142

7.9   Future System Design 143

8   Macromolecule Ranking on the DAP   144

8.1   Introduction 144

8.2   Background 144

8.3   The POSSUM System 145

     8.3.1   Representation of Secondary Structure Elements 145

     8.3.2   Calculation of Inter-Motif Similarity 147

8.4   Experimental Details 149

8.5	Serial Implementation of the Algorithm	150
8.6	Parallel Ranking Algorithm I	152
	8.6.1 Complexity of the Algorithm	153
	8.6.2 Results	154
8.7	Parallel Ranking Algorithm II	155
	8.7.1 Results	157
8.8	Bit-Level Algorithm	157
	8.8.1 Algorithm	157
	8.8.2 Implementation	159
	8.8.3 Results	160
8.9	Conclusions	160

9 Conclusions and Suggestions for Further Work — **169**

9.1	Conclusions	169
	9.1.1 Substructure Searching	169
	9.1.2 Macromolecule Ranking	171
	9.1.3 General Comments	172
9.2	Suggestions for Further Work	173
	9.2.1 Increasing the Performance of Existing Code	174
	9.2.2 Implementation of Algorithm IV	174
	9.2.3 Implementation of Screening Algorithms	174
	9.2.4 Other Applications of the Subgraph Isomorphism Algorithms	175

Bibliography — **174**

Appendix A Iterative Ullmann Code — **188**

Appendix B Query Structure Diagrams — **190**

Appendix C Abbreviations — **195**

Index — **196**

Chapter 1

Introduction

1.1 Objectives

The prolific growth in the use of the computer represents a striking development in modern times. Since the introduction of the digital computer in the early 1950's, for use within the scientific community, its use has become widespread. In this period there has been a constant demand for increased computing resources and this has been satisfied by a roughly ten-fold increase in the speed of computer arithmetic every five years. Such increases in performance have been sustained by technological developments in the manufacture of computers. The first generation of computers used valves which were both slow and unreliable; in the 1960's the second generation employed transistors which were faster and much smaller. The most startling improvements in processing speeds were seen after the advent of the third generation of integrated circuit technology, which allowed many thousands of transistors to be fabricated on the same piece of silicon. During this period all computers used an architecture proposed by Von Neumann, in which only a single item of data is processed at any time, so that sets of data are processed serially, one after the other. The performance of such computers has continued to increase in line with the advances in hardware design. Limiting factors, such as the speed of light, however, are now making further improvements in processing speed difficult. In order to sustain the continued demand for faster processing new computer architectures will need to be considered.

Parallel processing involves many processors co-operating in order to complete a task more quickly than a single processor. The 1980's were the decade of the parallel computer, and several machines designed explicitly for concurrent processing were introduced. Although many types of parallel computer now exist, no generally used methodology for the programming of parallel computers has become established. Instead, software engineers have designed parallel algorithms with a specific parallel computer in mind, in order to achieve the best performance. The purpose of this book is to analyse the suitability of a parallel processor known as the DAP (Distributed Array Processor) for the processing of databases of chemical information.

Computer-based chemical information systems provide facilities which allow the chemist to store and retrieve information pertaining to chemical structures. Such systems can store two-dimensional (2-D) chemical structure diagrams in a machine-readable format. An important retrieval facility offered is that of substructure searching in which all those structures which contain a user-defined query substructure are retrieved. Chemical structures may be regarded as mathematical graphs and the substructure search may be implemented by using techniques derived from graph theory. The graph-theoretical analogue of the substructure search problem is that of determining the presence of a subgraph isomorphism between two graphs. This theoretical problem belongs to a class of problems whose complexity has been studied extensively. The class of NP-complete problems, as they are known, are notoriously difficult to solve in a reasonable time and require large computing resources for all but the simplest of cases. This book describes the background to, and the implementation of, substructure searching on the DAP. Parallel subgraph isomorphism algorithms were designed for implementation on the DAP; the performance of the parallel system was then compared with that of a serial system executing on a fast serial processor, which was also designed and implemented as part of the research.

More recently, developments in molecular biology and biotechnology have led to the use of databases for the storage and retrieval of 3-D information about macromolecules such as proteins. There has been little published work describing retrieval systems for use with such databases; work has been done at Sheffield University to develop retrieval techniques for use with files of 3-D macromolecules. A novel system has been implemented which allows the retrieval of proteins which contain a user-specified substructural feature, referred to as a motif, in a similar fashion to 2-D substructure search, the difference being that the retrieved structures do not match the query exactly. In order for the user to identify the 'best' matches the retrieved structures are ranked in order of structural similarity with the query. The similarity measure used involves approximating the topography of a structure by a frequency distribution of the inter-point distances within the structure. It was found that although the ranking procedure produced satisfactory results, the response time was too long, because of the many distances that need to be calculated. Each distance calculation is independent and it was realised that many distances could be calculated simultaneously. For this reason the ranking algorithm is a good candidate for execution on the DAP. The implementation of the macromolecule ranking algorithm is the second application of the research described in this book. The diverse nature of the two applications provides a useful insight into the effectiveness of a parallel computer, such as the DAP, for the retrieval of information in chemical information systems.

1.2   Outline

Chapter 2 discusses the evolution and use of chemical information systems, in particular the different ways in which structures may be represented and various methods of retrieval. Chapter 3 introduces the subject of graph theory and describes its use in connection with retrieval in chemical information systems. Also covered is the subject of complexity, which is used to quantify the resources used by algorithms. In Chapter 4 parallel hardware and parallel algorithms are considered. Chapter

3

5 describes the hardware and software of the DAP in detail. The following three chapters report the experiments carried out. In Chapter 6 the Ullmann subgraph isomorphism algorithm is explained, together with descriptions of several parallel implementations of this algorithm and related performance data. The design and implementation of a substructure search system running entirely on the DAP is reported in Chapter 7. In Chapter 8 three novel algorithms for macromolecule ranking on the DAP are reported together with results. Finally, Chapter 9 contains an overview and summary of the results together with suggestions for further work.

In Appendix A a detailed description of the Ullmann algorithm is presented, together with a list of errors discovered during the course of this work. Appendix B contains a list of the structure diagrams corresponding to the query molecules used in the main substructure searching experiments described in Chapter 6. Lastly, a list of the most frequently used acronyms is given in Appendix C.

Chapter 2

The Storage of Chemical Information

2.1   The Evolution of Chemical Information Systems

An information system can be defined as having the following components [16]:

- The store of useful information.

- A series of techniques for adding material to and retrieving information from the store,upon demand.

- A group of people who operate the system.

- User(s).

Chemistry is described as 'the science of the elements and compounds and their laws of interaction and change resulting from substances in contact' [5]. Thus a chemical information system can be characterised by the storage of information concerning elements, compounds and their reactions. There follows a brief history of chemical information systems.

Since chemical compounds are of central importance in chemistry it is not surprising that their representation is of interest when considering chemical information

5

systems. Thus the representation of compounds for oral/textual communication will also be considered in this section. It is apparent from the above definition that an information system is not necessarily computer based. In fact chemists formed their own primitive chemical information systems long before the advent of the digital computer. Textbooks formed the earliest means of communication of information [15]. At this time chemists used trivial names for compounds, together with abbreviations which formed a shorthand set of symbols for the known chemicals. Later the scientific journal came into existence and this provided a system for the dissemination of the results of research in the same way it does today.

Following the discovery of the principles of modern structural theory there was a need for a more explicit representation of compounds; chemists began to use 2-D drawings showing atoms and their connections. These drawings were the first examples of the structure diagram. Although the structure diagram provided a means for graphic communication between chemists it was difficult to convey orally or by text. Nomenclature systems were developed at the same time which allowed oral/textual communication, but were tedious to use because of the large number of rules required for their correct use.

With the advent of computer technology, schemes were needed which allowed compounds to be retrieved according to different requirements [16]; the use of fragmentation codes was such a scheme. A fragmentation code comprises a set of substructural features, the presence or absence of which is used to characterise a compound [131]. Such fragmentation codes allow the retrieval of compounds in a variety of ways. Fragmentation codes were first used with punched cards; fragments were assigned fixed positions on a card and a punched hole indicated that the corresponding fragment was present in that compound. Although fragmentation codes allowed retrieval of structures according to substructural characteristics, identification of compounds could not be guaranteed since the coding scheme did not represent a structure completely. A representation was needed which completely characterised compounds and which could also be easily manipulated within a computer based system.

6

In answer to these requirements, several schemes were introduced which fell into the linear notation category. A linear notation is a coding mechanism which allows a chemical structure to be represented by a string of alphanumeric characters [131]. Only one such scheme has been used widely, the Wiswesser Line Formula [16]. The system was revised after various anomalies and deficiencies had been found and subsequently a listing of rules was produced called 'The Wiswesser Line Formula Chemical Notation' (WLN) [139]. SMILES (Simplified Molecular Input Line Entry System) [129] is an example of a more recent linear notation.

Although linear notations such as WLN provided unique and unambiguous representations of compounds, they were lacking in one respect. Because connections between atoms are not specified explicitly, searching for patterns of atoms (substructure search) is difficult. Text searching techniques can be used to search for certain substructural features, but conversion to a different representation is necessary for general substructure search. Another disadvantage of Wiswesser and other linear notations was the difficulty in manual coding/decoding of structures/notations. Without a graphics-based system for entry and display of results, interaction with a WLN-based system would be difficult for all but the most experienced of users. SMILES was designed specifically for chemists and seeks to overcome such difficulties by minimising the number of coding rules and providing an interactive computer-based system for structure entry.

In order to overcome such problems, topological schemes were introduced which specified atoms and their connections to other atoms directly. The connection table is the most successful topological representation. As its name suggests, the compound is represented by a list of atoms and the atoms to which they are bonded. The connection table can be considered a very general representation since it can be both unique, unambiguous, and also directly specifies the topology of a molecule. This generality enables full substructure searching to be performed, as well as easy conversion to other representations. The price paid for the generality is in storage requirements. Explicit representation of atoms and bonds requires more data to be stored than other representations, such as WLN, in which many structural features

7

are implicit.

The influence of computers in chemical information handling can be seen in the evolution of structure representation. Schemes such as fragmentation codes used in manual indexing systems were easily transferred onto the early computer systems of the time which used punched cards. Later, when improved backing storage became available, representation schemes were introduced, which were more suitable for handling by computers. The change-over, however, from manual to computerised chemical information systems was not completed in a single step. An example of this is the early use of computers by the Chemical Abstracts Service (CAS) [47].

The journal, Chemical Abstracts, started at the beginning of this century, its purpose being to abstract all fields of chemistry from the journals. The tremendous increase in production of information by chemists led to computers being used to assist in the production of Chemical Abstracts [15]. Later computer files containing indexes to compounds and subjects were produced. More recently on-line access was provided to the Chemical Structure Registry file, a file which now contains about 10 million [131] compounds. This file may be used to identify compounds and gain access to the abstract of the article which first reported the compound.

With the reduction in cost of main memory and backing storage, modern chemical information systems offer great possibilities. Large amounts of chemical data can be stored, and also detailed abstracts. Many large pharmaceutical organisations maintain their own in-house systems which contain chemical information generated within the organisation; and, recently chemical information systems have been expanded to accommodate chemical reaction data.

Thus one can appreciate the role of computers in the evolution of chemical information systems and their necessity for the continued improvement of chemical information handling.

2.2 Modern Day Use of Chemical Information Systems

An in-house chemical information system will contain much data about chemical compounds. Information likely to be stored is as follows [15]:

- Structure.

- Patent information.

- Physio-chemical data.

- Preparation details.

- Handling data.

When using such a system for the retrieval of chemical information the chemist issues a query to the system. Queries may be grouped according to information specified in the query and the resulting information which is retrieved. The first group contains queries in which structural information is specified; this information is used to identify compounds. Information associated with these compounds is then retrieved. Alternatively, non-structural information is specified in a query and this is used to identify compounds; such queries form the second query group. Examples of queries from both groups are given:

- How do I prepare a compound ? (structural).

- Retrieve all compounds with a structural feature (structural).

- Retrieve all compounds with a common physical property (non-structural).

Thus we have two entry points to such a system, structure and property. Most systems allow access to information from both entry points. Chemical Abstracts has indexes for structure and also properties [47]. The most important access point, however, is via the chemical structure since most questions asked of chemical information systems use this entry point [16]. Hence, emphasis has been placed on the

$$HS - CH_2 - CH_2 - Cl$$

Figure 2.1: Example of a simple structure diagram.

design of efficient structure representations for use in such systems. It is this access point that separates chemical information systems from other information systems. The structure diagram is closely related to the actual structure of a compound. Utilisation of the structure diagram as an entry point to chemical information systems allows designers to hide the complexity of systems and present a simple interface. Such simplicity enables chemists to use chemical information systems directly, without the need for intermediate parties with specialist knowledge. The importance of chemical structure representation necessitates a more detailed discussion of the commonly used coding schemes.

2.3 Representation of Compounds

2.3.1 The 2-D Structure Diagram

The structure diagram is a drawing which depicts atom-types and the bonds that link them with other atoms. Atoms are represented by their chemical symbols and bonds by lines linking symbols. An example of a structure diagram is shown in Figure 2.1. The structure diagram is used for communication of structural features between chemists. Other representation schemes seek to represent the structure diagram and thus the compound which the structure diagram represents.

Representation schemes can be characterised by their ability to fully define the compound which they represent. There follows an explanation of this classification.

2.3.2 Ambiguity in Representation

A representation scheme is said to be ambiguous if it is not possible to recreate the topology and labelling of atoms of a compound from its representation. Conversely,

10

an unambiguous representation allows the complete recreation of the topology and labelling of atoms.

Examples of unambiguous representations are:

- Systematic Nomenclature.

- Linear Notations.

- Connection Tables.

And ambiguous representations:

- Fragmentation Codes.

- Un-Systematic Nomenclature.

Another important characteristic of a representation is whether it produces a unique coding of a structure. A unique scheme will always produce the same representation when coding a structure. Ambiguity and uniqueness can be formally defined using functional analysis [61].

Let S be the set of all configurations of all types of atoms (not necessarily possible in the chemical sense), and T the set of all representations which can be generated from the set S. Then we can define a relation between the two sets S and T. The relation C is implemented by the algorithm used to code compounds. For C to be considered a function then, by definition, no element of S may by mapped by C onto more than one element of T:

$$\forall s_1, s_2 \text{ if } c(s_1) = t_1 \text{ and } c(s_2) = t_2$$

$$\text{then } s_1 = s_2 \Rightarrow t_1 = t_2$$

And this is the condition for uniqueness.

The function is injective if no member of T is the image under C of two distinct elements of S,

$$\forall t \in T, s_1, s_2 \in S$$

11

$$c(s_1) = t \text{ and } c(s_2) = t \Rightarrow s_1 = s_2$$

And this is the condition for non-ambiguity.

Thus a representation scheme is unique and unambiguous if the relation (coding algorithm) for mapping compounds onto codings is an injective (one-to-one) function.

As one might expect it is desirable to be able to define the exact structure of a compound from its encoded form, since this is necessary for substructure search. Consequently, unambiguous coding schemes are prevalent in chemical information systems, and ambiguous representation schemes will not be considered further.

2.3.3 Unambiguous Representation of Compounds

Systematic Nomenclature

Nomenclature schemes [29] represent a chemical compound as a name. Names which are not ambiguous are known as systematic nomenclature, ambiguous names are referred to as un-systematic nomenclature or trivial names. Chemical names which allow nearly complete re-creation of structure are called either semi-systematic or semi-trivial names.

Unique or canonical encoding of a structure cannot be accomplished algorithmically and is achieved intellectually using complex coding rules. This makes systematic nomenclature unsuitable for computer-based chemical information systems. Also, because the topology of a compound is not explicit in its representation, substructure search is impossible without conversion to a topological coding. Although unsuitable for computer-based systems, systematic nomenclature is often used in printed indexes [4].

Of the systematic nomenclature schemes, two are recognised internationally:

- International Union of Pure and Applied Chemistry (IUPAC) nomenclatures for organic chemistry and inorganic chemistry [4].

12

• The Chemical Abstracts Service Nomenclature.

Wisniewski reports the use of a system called AUTONOM [59,138] to automatically generate IUPAC-compatible names directly from structure diagrams; tests with random samples from the Beilstein database indicate that the system currently achieves a success rate of nearly two-thirds.

Linear Notations

A linear notation [38] is a coding mechanism which represents a chemical structure as a linear sequence of characters. Linear notations are thus similar to nomenclature systems; however, they are more explicit in structure representation than nomenclature. The predominant linear notation used in chemical information systems is the Wiswesser Line Notation (WLN) [18]. This system was developed by William J. Wiswesser in the early 1950's. WLN encodes chemical structures into a string of digits, letters and punctuation symbols; the available character set is doubled in practice by the use of a special character which acts as a shift key. A WLN representation can be thought of as a concatenation of substrings which represent functional types. In WLN common substructural features are often represented by a single symbol and this leads to compact representations. Inter-connections between atoms are not explicit in the coding but are implicit in the ordering of symbols within the representation. After some practice it is possible to derive a structure diagram from a WLN coding, but the reverse transformation, from structure diagram to WLN, is not as straightforward. This is because the meaning of symbols is context dependent, and thus the string of symbols before a symbol or group of symbols must be analysed in order to decode a substructural feature. For example, the WLN coding for the structure diagram shown in Figure 2.1 is SH2G.

WLN provides an unambiguous representation of a structure in nearly all cases. Furthermore, if the coding rules are followed exactly, the representation will also be unique. This makes WLN a good representation scheme for full structure matching. Although it is possible to search WLN strings for specific features [35], generalised

13

1	S	2(5)	
2	C	1(5)	3(5)
3	C	2(5)	4(5)
4	Cl	3(5)	

1	S	
2	C	1(5)
3	C	2(5)
4	Cl	3(5)

Redundant Form   Non-Redundant Form

Figure 2.2: Redundant and non-redundant connection tables.

substructure search is not possible when using a WLN representation since the topology of a compound is not represented explicitly. For this reason connection tables have become the preferred representation in chemical information systems.

Topological Representation: The Connection Table

Coding schemes which describe the topology of a chemical structure explicitly are referred to as topological representations. The most widely used topological coding scheme is the connection table [16]. When encoding a compound each atom in the structure is assigned a natural number. The connection table consists of a series of rows, one for each atom in the structure. Each row contains the atom number, the chemical symbol and a list of tuples specifying linkages to other atoms. Each tuple consists of an atom number and a bond number, where the bond number identifies the type of bond. An example of a connection table for the structure shown in Figure 2.1 is shown in Figure 2.2. The reader will note that hydrogen atoms and their bonds are not present in the connection table. This is true in general because the presence of hydrogen atoms can be deduced by considering un-satisfied valencies. In the first example each connection is specified twice, once for each atom connected by a bond. This type of connection table is termed 'redundant', since information about atom connections is replicated. In a non-redundant connection table each bond is cited only once. A connection table is unambiguous but not necessarily unique. This is because there are many possible labellings of the atoms in a structure; consequently a

14

single structure can be represented by many distinct connection tables. In order that connection tables can be matched directly a unique connection table representation must be derived.

As already stated, there are two general forms of connection table, redundant and non-redundant. The amount of space a non-redundant connection table occupies can be reduced if a single atom and its bonds are represented by a single symbol in the connection table. Also, each atom can be assumed to be connected to the previous one unless indicated. Such a scheme produces a 'notation-derived' connection table [16]. Although such a factorisation will minimise storage costs, it will prevent fast substructure search because atom connections are no longer explicit.

There are several advantages in using connection tables to represent structures. Coding and decoding of structures is a simple process because the rules for coding/decoding are trivial in comparison with other schemes such as WLN. This facilitates easy input and display of structures. Structures can be entered as redundant connection tables with arbitrary numbering of atoms; a unique coding can then be determined algorithmically. Fragment terms can be easily generated from a redundant connection table. Connection tables allow generalised substructure search since it is possible to examine the topology of a structure directly from its connection table.

Several factors have contributed to connection tables becoming widely used in chemical information systems. The flexibility of the representation together with modern demand for substructure search capabilities are perhaps the principal reasons for increased use. Also, falling secondary storage costs have made storage of large numbers of connection tables feasible. For these reasons connection tables will be the only form of representation considered from here onwards.

2.4 Retrieval in Structure-Based Chemical Information Systems

Structure based chemical information systems may offer several types of search facility:

- Structure search;

- substructure search;

- similarity search;

- generic search.

These will be described informally in this chapter, with formal graph-based descriptions and algorithms discussed in Chapter 3.

2.4.1 Structure Search

Structure search [131] involves matching a query structure exactly with each structure in the database. Thus the search is only successful if a database structure corresponding to the compound specified in the query is found. The chemist may use a structure search to ascertain property information about a compound in which he is interested, for example preparation details. Alternatively structure search is used in the process of compound registration. Registration involves the addition of a novel compound and associated data to a registry file. The registry file contains information about a compound generated or gathered within or on behalf of a company for exclusive use within that company. Each distinct compound is present once in the file; to determine whether a compound is novel, a structure search of the registry file must be performed. If there are no hits then the compound and its associated data are added to the file.

2.4.2 Substructure Search

Substructure search [15] is used to retrieve a set of compounds which are all related by a common substructural feature. The feature is specified and every structure in the file containing this substructure is retrieved. A structure is said to be a substructure of another structure if the first structure is contained within the second, irrespective of connections to the substructure and any other features present. There are several reasons why a chemist might execute a substructure search. If a chemist is interested in the chemistry of a substructure he may wish to retrieve all structures containing this substructure in order to compare properties. The chemist might also wish to examine the effects of a substructure in chemical reactions, and would therefore require a list of structures which shared the substructure in question.

Similarity search and the searching of files of generic structures are two less common types of search facility. Each may be useful in situations where structure/substructure searches cannot be used or produce inadequate results.

2.4.3 Similarity Search

The searching of a file for all those compounds with at least a certain degree of similarity to a target structure is called similarity search [131]. To achieve this, a quantitive measure of the similarity of two structures is required. The user specifies a structure and a level of similarity; and all the database structures with a similarity greater than or equal to this threshold are retrieved; the retrieved structures may then be listed order of decreasing similarity.

Similarity searching techniques are being investigated in response to structure and substructure search deficiencies in certain environments: consider a substructure search in which the substructure specified is too general. This will result in large numbers of structures being retrieved. The user that executed the search may find that many of the structures retrieved are irrelevant for his purposes. In this case a target structure could be specified and the ranking system would provide a means of identifying those compounds, structurally similar to the target, from a large set.

17

A chemist who wishes to identify compounds with similar structures to a query compound might also require a similarity search. This scenario might arise when a chemist is trying to find compounds which possess properties similar to a query compound. If the chemist does not know the substructural features responsible for these properties, substructure search could not be used and similarity search would be necessary. Also, there is considerable interest in correlation of property data with structural parameters: such associations are known as structure-activity relationships [15], in which structural similarity can be used to predict the activity of compounds.

The method used to quantify structural similarity is important; a bad similarity measure will lead to bad similarity search. Many methods of similarity measure exist and their effectiveness is an area of current interest [133]. One such method uses the concept of maximal common substructure (MCS).

Maximal Common Substructure

Given two compounds, the largest substructure common to both is known as the maximal common substructure. The concept of a MCS can be useful in two areas of retrieval in chemical information systems. Firstly, the MCS can be used as a similarity measure between two compounds, i.e., the larger the MCS, the greater the similarity. Thus MCS can be used in similarity search; this is unlikely, however, since the MCS problem is NP-complete and very demanding of computing resources.

Another area which utilises the concept of MCS is reaction indexing [15]. Chemical reactions can be characterised by the structural differences between the reactant and the product molecules [15]. The part of a reactant molecule unchanged by chemical reaction can be identified by a MCS algorithm; when the MCS has been found the structural differences between the reactant and product molecules can easily be identified. This approach to reaction indexing is attractive since it allows indexing of reactions by algorithm. As was noted earlier, computation of MCS is at present costly in terms of processor time. Algorithms have been developed, however, which

18

allow approximate calculation of MCS, and such approximations are close enough for indexing purposes [77,89]. At present the concept of MCS is not used widely elsewhere in chemical information systems because of the time complexity of MCS calculation.

2.4.4 Generic Structures and Search of Generic Structure Files

Structures in which alternatives are specified for particular parts of a molecule are referred to as generic structures or Markush formulae. Generic structures are characterised by the following properties [15,22]:

- The variable nature of substituent groups, usually expressed as a list of discrete alternative members;

- variable substitution patterns which specify the possible positions of attachment of these groups;

- the use of generic as well as specific nomenclature to express the chemical nature of substituent groups; and

- logical dependencies or exclusions which constrain the particular combinations of these groups.

The essence of generic structures is that they comprise an invariant part to which variant parts are attached at possibly variant locations. The attachment may be nested to an arbitrary number of levels [22]. Generic structures are commonly used for patent descriptions; this is because it is common for a patent application to cover a class of structurally related compounds. Such classes of compounds can be described by listing every member of the class. This is impractical, however, if the class is large and impossible if the class is infinite. Generic structure representation provides a compact way of describing such classes. Patent applications using generic structure representation have been accepted throughout this century [56].

19

In the past, fragmentation codes have been used to represent generic structures. Usually a fragmentation scheme is tailored to suit the requirements of the system's environment. Examples of such systems are:

- GREMAS (Generic Retrieval from Magnetic Tape Search) code [114] used by IDC (Internationale Dokumentationgesellschaft für Chemie) [56];

- RINGCODE and the New Chemical Code used by Derwent Publications for RINGDOC [25];

- DuPont code used by IFI/Plenum (Information for Industry) [19].

Fragmentation-based representations, however, suffer limitations as discussed earlier. There has been interest in representations based on modified connection tables [83,88,99]. Lynch's generic structure group has developed a generic structure description language, GENSAL [22], for encoding, input and display of generic structures from patents. Gensal has a strict syntax and provides a systematic basis for describing generic structures. A GENSAL description can be used to automatically generate a connection-table representation, known as an ECTR, of the generic structure.

When considering search of a file of generic structures, a query may be of one of the following types [22]:

- Specific structure.

- Substructure.

- Generic (Markush) structure.

A specific structure is a structure in which there are no variant parts, and therefore no substitution is permitted. A substructure query may or may not have generic parts, but must have at least one open valency, allowing it to be embedded in a full structure. Finally, a generic structure is defined as a structure in which at least

20

one part is invariant or fixed. A survey conducted by CAS [22] indicated that users would like to be able to search files of fixed and generic types using the above classes of query. The consequences of a system to allow such searches must, however, be considered carefully. A generic search of a generic file of structures might generate a large or infinite set of structures satisfying the query, and use large or infinite amounts of processor time. At present only a few systems exist which permit a generic search of a generic structure file; among these are MARPAT [51] and Markush DARC [119].

2.5 Structure and Substructure Search

2.5.1 Structure Search

Structure Matching

The way in which structure matching is performed depends upon whether the representation scheme used is unique or not. The connection table is in general a non-unique representation scheme, although it is possible to generate canonical codings of structures. If compounds are represented as non-unique connection tables then a compound might be represented by a number of different connection tables. The number of possible representations of a structure is the number of distinct labellings that can be assigned to the labels of a compound. Labels are assumed to be members of the set IN(the set of natural numbers). Therefore, for any structure with n atoms, there are $n!$ distinct label sets and $n!$ connection tables. To prove non-equivalence of a pair of compounds $n!$ connection tables need to be compared. Unless n is small, or the number of structures in the database is small, this method is infeasible.

Alternatively, compounds can be stored uniquely by generating a unique labelling of the graph representing a structure (this corresponds to finding a unique numbering of the atoms in a connection table). Connection tables can then be matched directly to determine if a pair of structures are equivalent. Finding an efficient algorithm to

21

generate unique connection tables is analogous to the coding problem [109] of graph theory, which is known to be a member of the complexity class NP. A commonly used scheme is the Morgan algorithm [98]. This algorithm assigns natural numbers to the vertices of a graph based on extended connectivity. These numbers are used to classify vertices, and classes are iteratively partitioned until each vertex belongs to a unique class or until no further partitioning of classes is possible (in which case an exhaustive method is used to produce a set of unique classes). The final vertex numberings are used to label the graph and produce a unique connection table.

Search of Structure Files

Files of connection table representations can be searched using serial or direct access techniques; this applies to both unique and non-unique collections of connection tables. If every compound in a file is compared with a query compound in a sequential fashion then a serial search has been made of the file. Implementation of serial search is trivial; starting with the first record in the file, and then the second, etc. comparisons are made with a query until a hit occurs or the end of the file is reached. Serially-organised files are easy to maintain because new records are simply added to the end of the file. Space overheads are low because no index mechanisms are required if access is always sequential. Serial files are not suited to systems in which real-time access is required to file records. This is because half of the file, on average, must be compared with a query to find a record. Therefore even if a canonical representation is used, enabling fast matching of individual structure records, the search will be slow. Serial access of a file of non-canonical representations will be even slower.

Direct access to structure files is usually implemented by hashing techniques [136]. A hashing function is used to produce a hash-code directly from a connection table. This code is then compared with hash-codes corresponding to each structure in the file. A serial search is then made of the few structures having the same hash code. Hashing techniques are generally used with unique connection tables although they may also be used with files of non-unique connection tables [24]. At present most structure-based search systems use direct access, since this is the only way of ensuring good real-time performance.

2.5.2 Substructure Search

Introduction

The techniques described above allow searches of large files of structures in times which are acceptable for real-time use. The same methods cannot be used in practical substructure search systems. If a hashing procedure were used to locate structures containing a query structure, unacceptable amounts of storage would be needed. This is because the set of all possible substructures of each structure would need to be stored in order to allow search for arbitrary patterns of atoms and bonds. Such an approach would allow fast substructure search but would be NP-complete in terms of storage requirements.

Real-time substructure search can be achieved by an alternative approach. The search is divided into two stages. The first stage seeks to eliminate structures which cannot possibly contain a query structure. This part of the search is called the screen search [15]. Remaining structures are then passed onto the second stage of the search. A computationally expensive algorithm is used to try to establish a subgraph isomorphism between the database and query structures. This stage is referred to as the atom-by-atom match. The screening stage of the search is inexpensive compared with the atom-by-atom match; therefore it is important to minimise the number of false-drops passed to the second stage of the search. There follows a detailed discussion of the two-stage process used to effect substructure search.

Screening Systems

Representation Of Screens A compound can be characterised by a set of screens. The screen set provides an approximate description of a compound for screening purposes, rather than the exact representation which is required for atom-by-atom matching. The screen set associated with a compound is determined by the presence or absence of a number of fragments in much the same way that a compound is represented in a fragment-based system. A bit-string is used to represent the screens. There are two common methods used to code a compound. In the first approach each bit position corresponds to a unique fragment from the screen set. If the bit is set then this indicates the presence of the fragment associated with this bit position. A second method involves each fragment being associated with several bit positions [50]. A screen set is calculated by taking the logical OR of the bit patterns corresponding to every fragment present in the compound to be coded. This method reduces the amount of storage required by screens, since a screen set can be characterised by a shorter length bit-string. This reduction in length is achieved at the expense of a reduction in the effectiveness of the screens. The implementation of a screening search is independent of the system used to code screens.

Choice Of Screen Sets The term 'screen-out' is used to measure the proportion of database structures which are eliminated from further consideration by the screening stage of substructure search. Screen-out is affected by many parameters, the most important of which is the screen set chosen to characterise compounds. A poorly chosen screen set will result in a large number of compounds being passed to the compute-intensive atom-by-atom matching stage. In contrast an optimal screen set will maximise screen-out, and Lynch points out that under the right conditions a single structure can be distinguished from a file of one million compounds using just 20 screens [16]. It should be noted that a good screen set will not necessarily contain chemically significant fragments. Mooers [97] was the first to put forward a methodology for choosing an optimal screen set. Later Lynch [87] verified this methodology using an empirical approach. A theoretical explanation of the results obtained by Mooers and Lynch was then presented by Hodes [74]. Following is a summary of the theoretical and practical considerations that affect the choice of screen set.

A consistently high screen-out over a broad range of queries is achieved if each screen is present in exactly half of the file structures and each screen is independent. It is difficult to find a set of screens that satisfy these properties in a chemical structure file. Only a small number of fragments will exist in half of the structures in a file since there is a large distribution of structural features. The nature of this distribution is explained if the variation of occurrence of the elements is considered: carbon(72%), oxygen(14%), nitrogen(7%), others(7%) [15]. Furthermore, any screen set chosen is likely to contain fragments which are not independent of one another. A further obstacle in achieving high screen-out is the size of query structures. The average size of a query will be smaller than the average database structure. The query structure will consequently contain fewer screens. Frequency of occurrence of screens must also be considered. A screen set which contains screens which occur at similar frequencies in the structure file will increase screen-out. This can be explained by considering two screens with disparate frequencies. The first has a high frequency of occurrence, and is likely to occur in queries. Although this screen will be used often it will not eliminate many structures. Conversely, a screen with a low frequency of occurrence will eliminate many structures but will not be utilised often. To achieve consistently good screen-out with a broad range of queries a screen set with a homogeneous set of occurrence frequencies is required.

There are also practical considerations which influence the choice of a screen set. Large specific screens are required when searching a file of homogeneous structures. Small generic screens will suffice in a heterogeneous structure file. The probable nature of queries must also be considered; if many substructure searches are to take place then it is necessary to maximise screen-out by maintaining a large screen set. If the system is to be used mainly for structure search, however, a large screen set will be unnecessary. The size of the structure file will also affect the nature of the screen set. It is imperative to achieve high screen-out when dealing with a large file

25

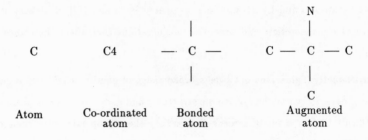

Figure 2.3: Atom-centred fragments.

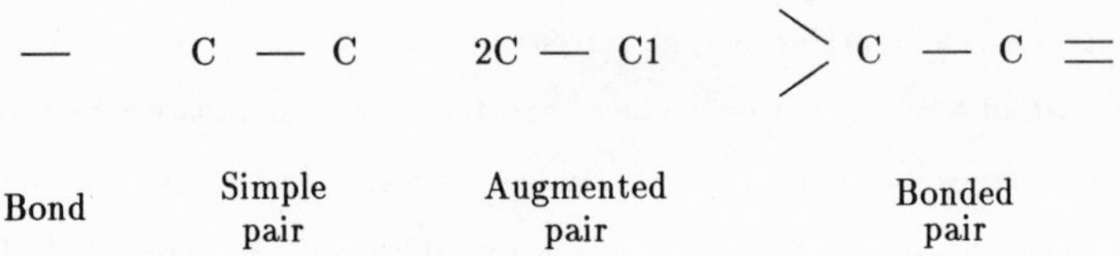

Figure 2.4: Bond-centred fragments.

but not so important if the structure file is small. Finally, file organisation must be considered: inverted files will suffer a performance degradation if a large screen set is used; this will be explained later.

Algorithmic Screen Generation   It should be apparent from the above discussion that a method of generating fragments which occur with varying frequencies is required in order to choose an optimal screen set for a collection of structures. The Sheffield method [11] algorithmically generates a hierarchy of atom and bond centred fragments as shown in Figures 2.3 and 2.4. Fragments can be grown until the required frequency of occurrence is achieved. A similar method of screen generation was developed at the Walter Reed Army Institute of Research (WRAIR). The

26

WRAIR approach grows fragments in a linear fashion, this resulting in fragments consisting of chains of atoms and bonds [50,100]. Both methods can be employed to create screen sets automatically from large files of connection tables.

Implementation Of Screen Search The method of structure file organisation influences how a screen search is effected. A bit-map can be constructed in which each row corresponds to the screens of a file compound, and each column corresponds to a specific structure screen. A screen search may be executed in one of two ways. If a query screen is compared with each row of the bit-map in a sequential fashion then this is equivalent to search of a serial file. Inverted file search is achieved by intersecting columns of the bit-map corresponding to the screens present in a query. The bit-map approach allows both serial and inverted file search. In a serial file each screen would be linked to a connection table, whereas screens and connection tables would be stored separately in an inverted file.

Atom-By-Atom Search

The process of finding a subgraph of a database graph which is isomorphic to a query graph can be implemented as a search. The search space can be formulated as a tree where the vertices at each level correspond to query atoms, and where the edges connecting nodes between levels correspond to mappings from query to database vertices. Thus each vertex will have children corresponding to all possible mappings of a query atom to a database atom. The number of branches in the tree will be the total number of possible mappings of the query graph onto database subgraphs. A search of the resulting tree can be effected using search strategies such as depth-first search or breadth-first search (depending on whether a single subgraph isomorphism or all subgraph isomorphisms are required). An isomorphism test would be applied after the creation of each branch of the tree.

Exhaustive Search The simplest approach is to generate all possible subgraphs (of the same size as the query graph) of the database graph and test for

27

isomorphism in each case. This technique will involve many operations as one can easily show:

The number of ways of selecting Q atoms from N atoms is

$$\frac{N!}{Q!(N-Q)!}$$

and this is the number of database subgraphs of the same size as the query graph. Each of these graphs can be labelled in $Q!$ different ways, therefore we have

$$\frac{N!Q!}{Q!(N-Q)!} = \frac{N!}{(N-Q)!}$$

graphs to test for isomorphism. This can be implemented as a search of the tree search space described. In the worst case the search will examine the above number of branches, and test for isomorphism in each case. This method will be prohibitively time consuming for all but the smallest structures. Practical search techniques seek to prune the tree and thus reduce search space.

Non-Exhaustive Techniques In order to reduce the search space it is necessary to identify branches which cannot possibly lead to isomorphism. This can be achieved by characterisation of graph vertices. Characterisation can be accomplished using either topological information and/or extra information present in the connection table, such as atom and bond-types. If a mapping between two atoms is to be considered further they must both exhibit the same features, for example, atom-type. Using this technique, non-correspondence between query and database atoms can be established quickly in the same way that screening is used to eliminate complete graphs from atom-by-atom search. Once such a non-correspondence is identified, the branch starting at the vertex associated with the query atom can be removed from the search space. If many branches can be pruned in this way then the work involved in finding an isomorphic subgraph can be greatly reduced. If, at any level, all branches are pruned then there can exist no isomorphism between the query graph and a database subgraph.

There exist several strategies for reducing the search space. Each of these methods remove branches from the search tree in the way described. The method used to

28

prune branches characterises an atom-by-atom search technique. It should be noted that unless an atom-by-atom search shows that a subgraph isomorphism does not exist, the search space left after pruning will always need to be searched. Following is a discussion of pruning methods.

Backtrack Search This technique prunes the search tree during depth-first search. Branches are pruned using atom connectivity information, bond-types and atom-types. This method is referred to as *backtrack* search since, after a mismatch is identified, the depth-first search proceeds to undo the previous atom assignment.

The following techniques seek to place atoms from the query and database structures in classes according to their characteristics. Classes are algorithmically partitioned to reduce the members present. This partitioning of classes is based upon further examination of atom characteristics. The process corresponds to the elimination of mappings between query and database atoms and consequently the reduction of search space.

Set Reduction Set reduction [92] uses set intersection procedures to partition sets. A number of sets, each corresponding to distinct atom properties, are constructed. Separate sets exist for query and database structures. For example, there might be a set containing the labels of all carbon atoms and also a set for those atoms which are connected to a neighbour by a double bond. Further sets are produced by intersecting existing sets. Thus, the set consisting of carbon atoms connected to a neighbour by a double bond could be found by intersecting the sets described previously. The process continues until the number of possible mappings between database and query atoms is reduced to a reasonable level or no further set partitioning can be achieved. If at any stage during set reduction the number of members of a database set is less than the number of members of the corresponding query set, indicating that a subgraph isomorphism cannot exist, the process then terminates.

Relaxation Relaxation techniques are used for pattern matching. An initial correspondence is made between two patterns by establishing mappings between individual components of each pattern according to some attributes of the components. The initial mapping is iteratively improved by seeking to extend the local similarity of components. The process continues until a threshold is reached, no further refinement of mappings is possible, or until it is found that mappings for one or more components cannot exist. The relaxation technique is used to label vertices in image processing [84], and for approximate solution of systems of equations in numerical analysis, as well as other areas in which pattern recognition is necessary. Von Scholley has developed an algorithm for use with generic structures [126]. Here, the specific structure version of the algorithm is described.

Firstly, initial mappings of query vertices to database vertices are made. The mappings are based on node and bond-type. Initially, a query vertex is only mapped to database vertices which have the same atom-type and bond-types. It is possible that a single database atom will be mapped to a number of query atoms; this group of atoms is referred to as a label set. The label sets of all the database atoms combine to form the label map of the structure. The refinement process attempts to eliminate labels from the label map of a structure. This continues until no further refinement is possible or until one or more query atoms can no longer be mapped to any database atom, in which case a mismatch is established. The refinement procedure uses three different techniques to refine atom mappings:

- Neighbour examination: if a database atom is mapped to a query atom then there must exist mappings between all of the respective neighbours. If not, the query atom is removed from the label set (this is the Ullmann heuristic, see Chapter 6).

- Atom connectivity: if a database atom has degree less than a query atom then the query atom label is removed from the database atom label set. The degree of a database atom will be decremented if, during refinement, a neighbour's label set becomes empty.

- Assignment: if a query label occurs only once in a label map then it can be assigned to a single database atom. Any other labels in the label set can be removed.

The above algorithm will not necessarily produce unique mappings between every query atom and a single database atom; however, the search space will usually be reduced.

2.6 Current Chemical Information Systems

2.6.1 Introduction

Current chemical information systems can be separated into two categories according to the users of the system. Firstly, there are systems such as CAS ONLINE which allow the public to access its files. Also, there are systems for exclusive in-house use. Among the reasons for storing data exclusively within an organisation might be the need for confidentiality of structures and data generated within the organisation, and the extra functionality provided by customised software. Although several organisations have developed such systems for their own use, such as COUSIN (CompOund Search Information System [64,65]) and MMS [81], such an approach will be uneconomical for small organisations requiring complex software. Many organisations have chosen to tailor general purpose software to their own needs. Such a system is MACCS (Molecular Access Program).

Both CAS ONLINE and MACCS will be described. The former as an example of a chemical information system available for public use and the latter as an example of an in-house system.

2.6.2 The CAS ONLINE Search System

The prototype CAS ONLINE search system [46] was introduced in 1980 to search the CAS Registry File. The original version of the system allowed screen-only searching

31

of a fraction of the file. Structures in the file were stored in connection table format to allow subsequent substructure searching to be implemented. Later versions of the system allowed structure and substructure search of the whole file and also the use of generic queries.

Each structure in the system is represented by a Unique Chemical Registry Record (UCRR). This includes a bond-explicit connection table as well as other information such as stereochemical characteristics, registry number, molecular formula and transaction data. Structure input to the system may be performed using either a character-based terminal or a graphics terminal. The use of graphics in the system was an early requirement since it was thought that such a user interface would increase the functionality of the system from a user's point of view.

Before 1988 the CAS Registry File was based upon a serial form of file organisation. Such an organisation greatly simplifies file maintenance. But, as detailed earlier (see Section 2.5.2), the scheme suffers from disadvantages including reduced search speed due to I/O overheads. In order to provide a reasonable response time the file was partitioned into segments with a pair of mini-computers responsible for searching each segment. One performed the screening operations and passed possible hits to the other for the atom-by-atom stage of the search. Processing was overlapped and thus some parallelism was achieved. Also, since the pairs of mini-computers searched the segmented file together, further parallelism was achieved; this type of parallelism is referred to as database parallelism. An IBM 3084 mainframe, known as the 'front-end controller' handled query input and the supervision of the search processors which were collectively known as the 'search machine'. Pairs of mini-computers could be added to the search machine to increase the search speed. In practice this was done after the file increased by about 700,000 structures, in order to maintain a constant response time as the file grew larger.

A new hardware and software configuration was introduced in 1988, called the 'Search Engine Complex'. The new system was designed in response to new requirements and also the availability of new hardware and software. It was intended that the new system support combined structural and textual searches as well as the ability to store generic structures. Parallelism is retained but is now inherent in the inverted file search process as well as the database parallelism described earlier. The software implementing the inverted file search is written in UNIX C which makes the system portable. Removable disks have been replaced by fixed drives which are smaller and offer better price/performance.

The search engine complex is generally quicker and more flexible than the search machine. The use of inverted files has decreased the average response time from 5-6 minutes to 1-2 minutes. Also, the addition of new screens is now a simpler task, because of the file organisation.

2.6.3   MACCS Molecular Access Program

MACCS [36,105] is a modular software package for input, storage and retrieval of chemical information. Prominent in this package is the use of graphics in all stages of structure input and output. The system is interactive and menu driven and offers sophisticated structure entry and display features. Stereochemical information can be described in a completely defined molecule. Structure and substructure searching are supported. In-house implementations of the system have been documented by several companies [9,10,20,21,79,86].

As the demand for data about the three-dimensional characteristics of molecules has risen an enhanced version of MACCS has been introduced, called MACCS-3D [96]. Among the design goals of MACCS-3D were [135]:

1. To provide a flexible set of geometric search tools for searching 3-D molecules.

2. To provide an open architecture for communication with other chemical information systems.

3. To create an interactive program which is capable of being used by a large body of chemists unfamiliar with computational chemistry.

33

The database is divided into two parts: the first stores the 2-D structure and other information associated with the molecule; the second part of the database contains 3-D models of molecules. 2-D molecules are linked with 3-D molecules by a one-to-many mapping, allowing several 3-D conformations to be stored for a single structure.

Many of the search options possible in MACCS will work in a consistent manner in the 3-D domain. For structure and substructure search the user specifies geometric constraints; MACCS-3D determines the equivalence of two 3-D models by determining the best 3-D fit required to overlay one model onto the other.

2.6.4   The New Generation of Chemical Information Systems

Greatly increased performance for given cost of micro-processors and memory devices has made micro-based chemical information systems attractive. Also, high quality graphics devices have become available due to new technology. This, coupled with the large storage capabilities of CD-ROM and the potential this medium has for storing graphics with text should lead to an increase in the number of micro-based chemical information systems [122]. Also, the development of parallel hardware and software should, in the long term, allow substructure searching of files to be performed without expensive processors.

Chapter 3

Graph Theory and Complexity

3.1 Graph Theory

3.1.1 Introduction

The term graph is used in two senses in mathematics. The first sense is that of a graph of a function, for example:

$$f : A \mapsto B$$

then

$$\text{graph}(f) = \{(a,b) \mid a \in A, b = f(a)\}$$

Subsets of such a set can be displayed in the Cartesian co-ordinate framework. Alternatively, the term graph is used to define an abstract mathematical entity. A graph in this sense can be informally described as a set of vertices or points and a set of edges or arcs which link pairs of vertices. Graph theory [134] is the branch of pure mathematics concerned with graphs of this type. If an object or situation in the real world can be abstracted as a graph, then graph theory can be used to analyse the problem domain. Modern day uses of graph theory include finger print identification [75], electrical network design and optimisation [123] and image processing [23,55]. Unlike many other areas of pure mathematics graph theory was invented to solve a real world problem, the 'Konigsberg Walk' [12].

Graph theory is relevant to structural chemistry because a compound for which the chemical structure is known can be represented as a graph. The atoms and bonds of a compound are represented by the vertices and edges of a graph. The abstract representation of chemical compounds in this way allows graph theory to be used as a tool in the analysis of structure-based chemical information systems. Before elucidating further, a few formal definitions are presented.

3.1.2 Formal Definitions

A graph can be characterised, in the general sense, by the attributes of its vertices and edges. A graph which has a direction associated with each edge is said to be a directed graph. Labelled graphs have a unique descriptor associated with each vertex; similarly, a weighted graph has a unique descriptor associated with each edge. Chemical structures can be represented by labelled, weighted, un-directed graphs. Therefore we will restrict our attention to this type of graph.

Definition:

An un-directed graph $G(V, E)$ is specified by two sets, where:

V is a set of vertices,

E is a set of edges connecting vertices $E \subset V \times V$.

An un-labelled, un-weighted, un-directed graph may be represented by an incidence or adjacency matrix. This is a two-dimensional symmetric Boolean array in which rows and columns correspond to nodes in the graph. A TRUE element indicates that a pair of nodes are connected, while a FALSE element represents an unconnected pair. An example of such a graph together with its associated adjacency matrix representation is shown in Figure 3.1.

3.1.3 Graph Theory and the Connection Table

A graph can be used to represent the topology of a chemical structure. Similarly, the topography (3-D characteristics) of a compound can be represented as an un-

36

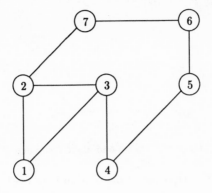

$V = \{1,2,3,4,5,6,7\}$
$E = \{\{1,2\},\{1,3\},\{2,3\},\{2,7\},\{3,4\},\{4,5\},\{5,6\},\{6,7\}\}$

$$\text{Adjacency matrix} = \begin{vmatrix} 1 & 1 & 1 & 0 & 0 & 0 & 0 \\ 1 & 1 & 1 & 0 & 0 & 0 & 1 \\ 1 & 1 & 1 & 1 & 0 & 0 & 0 \\ 0 & 0 & 1 & 1 & 1 & 0 & 0 \\ 0 & 0 & 0 & 1 & 1 & 1 & 0 \\ 0 & 0 & 0 & 0 & 1 & 1 & 1 \\ 0 & 1 & 0 & 0 & 0 & 1 & 1 \end{vmatrix}$$

Figure 3.1: A simple graph and adjacency matrix.

directed,labelled, weighted graph, where the Euclidean distances between each atom and every other atom are stored as the bond labels. A connection table represents a labelled, weighted, un-directed graph. In a connection table, edges correspond to atom bonds which have no direction, therefore edges are un-directed.

Often in the literature, atomic numbers are said to be vertex labels and bond-types edge labels. If we are to use the terminology of graph theory, this cannot be true. In the graph theoretic sense a graph is said to be labelled/weighted if and only if each vertex/edge has a unique label associated with it. Atomic numbers and bond-types cannot, therefore, be considered as labels since a compound will typically have one or more atoms/bonds of the same type.

$$O \xrightarrow{\ \beta\ } C \xrightarrow{\ \alpha\ } C \xrightarrow{\ \alpha\ } N$$

1	O	2(β)	
2	C	1(β)	3(α)
3	C	2(α)	4(α)
4	N	3(α)	

Figure 3.2: Structure diagram and associated connection table.

It is possible to find identifiers which can be considered to be labels. In a connection table, atoms are numbered uniquely and the number associated with an atom may be considered as its vertex label. Edges may be labelled by associating with each bond a set $\{a,b\}$ where a and b are the vertex labels of the atoms linked by the bond. A connection table also contains extra information, atom identifiers and bond-types. Implicit in a connection table, therefore, are two mappings. A mapping from vertex label to atomic number, and a mapping from edge label to bond type. An example will help to clarify the terms used.

Consider the connection table shown in Figure 3.2; this may be represented using the following notation:

$G = (V, E)$ where

$V = \{1, 2, 3, 4\}$

$E = \{\{1, 2\}, \{2, 3\}, \{3, 4\}\}$

In order to specify the information relating labels to atom and bond-types, two extra sets must be defined together with associated mappings:

$A = \{O, C, N\}$ (atom types)

$B = \{\alpha, \beta\}$ (bond types)

The relation $X : V \mapsto A$ maps node labels to atom types and is defined by the table:

38

v	X(v)
1	O
2	C
3	C
4	N

Similarly, the relation $Y : E \mapsto B$ maps edge labels to bond types and is defined:

e	Y(e)
{1,2}	β
{2,3}	α
{3,4}	α

It is considered necessary to provide such a mathematical treatment of the connection table representation of a molecule in order to make clear how a chemical structure may be treated as a graph. Such a description also serves to remove anomalies caused by different interpretations of graph terminology.

3.1.4 Structure/Substructure Search and Graph Theory

Definition: Graph Isomorphism

A graph G_1 is isomorphic to a graph G_2 if there exists a one-to-one mapping ϕ, called an isomorphism, from $V(G_1)$ onto $V(G_2)$ such that ϕ preserves adjacency and non-adjacency. In a graph G, the edges $v_1, v_2 \in V(G)$ are said to be adjacent if $\{v_1, v_2\} \in E(G)$.
Formally: given any $v_1, v_2 \in V(G_1)$

$$\{v_1, v_2\} \in E(G_1) \Leftrightarrow \{\phi(v_1), \phi(v_2)\} \in E(G_2).$$

Informally, two graphs are isomorphic if they share the same topology. Finding a graph which is isomorphic to a specified graph is analogous to matching chemical structures. For two chemical structures to be considered equivalent, not only must the topology of both match, but also the atom and bond-types. Graph isomorphism is a weaker equivalence, therefore, than chemical structure equivalence. It is not enough for a pair of connection tables to represent isomorphic graphs, since the bond-types and atom-types associated with the vertex and edge labels must also be identical for chemical equivalence.

Definition: Subgraph

A graph $G_1 = (V_1, E_1)$ is a subgraph of $G_2 = (V_2, E_2)$

$$\text{if } V_1 \subseteq V_2 \text{ and } E_1 \subseteq E_2$$

The problem of determining whether a given structure is a substructure of another structure corresponds to the problem of subgraph isomorphism.

Definition: Subgraph Isomorphism

Given two graphs $G_1 = (V_1, E_1)$ and $G_2 = (V_2, E_2)$ there must exist $V\prime \subseteq V_1, E\prime \subseteq E_1$ such that

$$\mid V\prime \mid = \mid V_2 \mid, \mid E\prime \mid = \mid E_2 \mid$$

and a one-to-one function $\phi : V_2 \mapsto V\prime$ such that

$$\{u, v\} \in E_2 \Leftrightarrow \{\phi(u), \phi(v)\} \in E\prime$$

As with graph isomorphism we must also match atom and bond-types in order to satisfy chemical substructure equivalence.

3.2 Complexity

3.2.1 Introduction

When trying to solve a problem by algorithm, it is important that the problem is first analysed because it may be that the problem is either inherently intractable or unsolvable by algorithm. Since the discovery of such intractable or undecidable problems, ways of analysing the 'difficulty' of problems have been introduced. Complexity analysis seeks to quantify the resources required by algorithms, and to use this information to group problems solvable by algorithm into classes. Problems can often be proved equivalent in terms of complexity. Such a classification allows algorithm designers to predict whether an efficient solution to a problem is likely to be found before an algorithm is constructed. For a more detailed account of complexity, it is necessary to introduce some terminology and notation:

A problem will be a general question for which an answer is to be provided. The problem will be characterised by:

- a set of parameters;

- a set of properties that a solution to the problem is required to satisfy.

An instance of the problem will be specified by providing a value for each problem parameter. An algorithm is said to solve a problem if a solution can be found for every instance of the problem. The efficiency of an algorithm is related to its requirements of computing resources. The central processor of a computer is considered to be the most important resource because speed of execution is regarded as the primary factor in deciding whether an algorithm is useful. The most efficient algorithm, therefore, is considered to be the one which can find a solution to a problem in the shortest time. In order to provide a general description of how efficient an algorithm is, we specify the time taken to arrive at a solution in terms of the size of the problem. Here, size is the amount of input data required to define an instance of the problem.

Definition: The time complexity function for an algorithm is a function which maps the input length to the largest amount of time required by the algorithm to solve a problem instance of that size. The domain of the function is the set of all possible input lengths.

The efficiency of an algorithm with respect to a problem is characterised by its time complexity function. In order to classify algorithms by their time complexity functions, further notation is necessary:

Definition: Assume f and g are functions, then $O(f(n))$ (read order at most $f(n)$ or order $f(n)$) is the set of all $g(n)$ such that there exist constants $c, n_0 > 0$

such that

$$| g(n) | \leq cf(n) \; \forall \; n > n_0.$$

3.2.2 Complexity Classes

Informally, algorithms which have time complexity functions belonging to the set $O(f(n))$, where $f(n)$ is a polynomial, are considered efficient or reasonable. Conversely, if $f(n)$ is exponential, the algorithm is considered inefficient or unreasonable. In order to make any formal classification of algorithm efficiency, a model of computation must be presented. A convenient model of a general purpose serial computer is the deterministic Turing machine [61]. Another model of computation used is the non-deterministic Turing machine [61], in which the machine's action is not fixed at every stage in a computation. Where a choice of action occurs, all possible actions are pursued in parallel. It has not yet been proved that a non-deterministic Turing machine is more powerful than the deterministic model. Certain problems exist, however, for which deterministic algorithms with the same complexity as existing non-deterministic algorithms have not been found. The two models of computation are, therefore, useful in the classification of problems.

Definition: The class of all problems solvable in polynomial time by a deterministic Turing machine is P (P stands for polynomial time).

Definition: The class of all problems solvable in polynomial time by a non-deterministic Turing machine is NP (Non-deterministic Polynomial time).

It is clear that P$\subseteq$NP, since a non-deterministic Turing machine can easily simulate a deterministic Turing machine. The question P $=$ NP ? however, is one of the foremost problems of theoretical computer science and pure mathematics. Problems exist for which algorithms in NP have been found but not in P, but this is not enough to prove P $\neq$ NP; in such cases it must be proved that no algorithm belonging to P can ever exist.

3.2.3 NP Completeness

It has been shown that one problem belonging to NP, the satisfiability problem, has the property that all other problems in NP can be reduced to it in polynomial time [34]. This problem may be regarded, therefore, as the 'hardest' member of NP since, if any problem in NP is proved not to belong to P, then this would also prove that the satisfiability problem does not belong to P. Also, if the satisfiability problem is found to belong to P, then P = NP would be a corollary. It has also been shown that there exist other members of NP which share the properties of the satisfiability problem in that they can be considered the hardest problems of NP. These problems are equivalent in terms of complexity and form an equivalence class known as the class of NP-Complete problems [58].

3.2.4 Hard Problems in Graph Theory

Many graph-theoretical problems have been shown to be NP-complete. Included in these are the problems of finding a subgraph isomorphism (of which graph isomorphism is a special case but still a member of NP) and finding the maximal common subgraph of two graphs [58]. This would suggest that structure and substructure searching is infeasible unless the structures concerned are very small. In the real world, compounds are not small enough to allow rapid search by inefficient algorithms. Therefore algorithms must be found that allow structure and substructure search of chemical databases to be performed in reasonable time. As discussed earlier, chemical structure search may be executed efficiently by the use of hashing techniques; however, this does not apply to substructure search.

3.2.5 Dealing With NP-Complete Problems

When faced with an NP-complete problem, one can pursue one of three alternative courses. The first is to try to find an efficient algorithm. Although it has not been proved that efficient algorithms do not exist for NP-complete problems, they have

43

never been found, and it is widely felt amongst mathematicians that $P \neq NP$ is likely. This approach, therefore, would probably end in failure since, if such an algorithm was found, $P = NP$ would be a corollary.

A second approach, which is only suitable for optimisation problems, is to relax the solution conditions. If nearly optimal solutions of a problem are acceptable, then it may be possible to find an efficient algorithm to do so. Such algorithms use heuristic methods to reach a good solution; however, good solutions cannot usually be guaranteed for all problem instances [58]. Efficient approximation algorithms which guarantee performance levels have been found for many NP-complete problems, including the classic travelling salesman problem [113]. Although this approach is only suitable for optimisation problems, many problems arising in the real world can be formulated in this way.

Finally, if a problem cannot be phrased as an optimisation task, or exact/optimal solutions are always required, a tree-pruning approach can be adopted. Partial solutions are generated within a search tree and partial solutions which can never be extended to full solutions are identified. Such partial solutions can be eliminated from the search space. If a good heuristic method is used to identify un-fruitful partial solutions high in the search tree, large amounts of search space can be removed and solutions can be found quickly. Tree pruning methods cannot be guaranteed to work efficiently for all problem instances, but by fine-tuning the heuristic part of the search algorithm, quick solutions can be found for a subset of problem instances.

Structure/substructure matching requires exact solutions of the graph/subgraph isomorphism problem. If inexact matches of compounds are allowed, false-drops might be generated; moreover the number of false-drops would be unpredictable and performance would be similar to fragment based search systems. If reasonable response times in a structure/substructure search system are to be achieved, tree-pruning techniques need to be utilised.

44

3.3 Conclusion

Various types of search have been described in Chapter 2, in particular the implementation of structure and substructure search. While graph isomorphism is a member of the complexity class NP, the analogous problem of structure matching can be performed in reasonable time, in most cases by the use of hashing techniques. This is not so, however, for substructure search. The problem of finding a subgraph isomorphism between two general graphs is NP-complete. Although techniques can be employed which reduce the combinatorial search space generated by the problem, this does not result in fast search times as with structure searching. Rather than pursue software techniques (which seem unlikely to be successful since it is widely felt that P$\neq$NP), another approach is to try to exploit any parallelism inherent in the problem. Since the subgraph isomorphism problem belongs to NP, it can be predicted that the problem could be solved in polynomial time if an infinitely parallel computer was available. Thus, it seems likely that a computer of finite parallelism would solve the problem more quickly than a serial computer, in the general sense. One cannot be sure of this since it has not yet been proved that a non-deterministic (parallel) Turing machine is inherently more powerful than a deterministic (serial) Turing machine.

Since software techniques have been widely explored already, it seems reasonable to try to find a fast solution to the substructure search problem in the largely uncharted domain of parallel processing. The next chapter describes how parallelism may be incorporated into the design of a computer and also gives examples of parallel processors that have already been implemented.

Chapter 4

Parallel Processing

4.1 Models of Computation

4.1.1 The Von Neumann Model of Computation

The Von Neumann model of computation [103] has formed the basis of most computers from Babbage's difference engine [108] to the fourth generation machines in current use. In this model of computer organisation, a program is stored in a memory and a single control unit is used to interpret instructions fetched from the memory, one after the other. A sequence control register determines the next instruction to be executed. Each instruction acts upon a single item of data; data being passed between instructions by means of references to shared memory locations. This model was adopted for early computers because the cost of arithmetic units was high relative to memory and, therefore, replication of arithmetic units within a single computer was costly.

When the first high level languages were developed for the second generation machines, these reflected the underlying hardware and were explicitly sequential (that is, apart from LISP, which was developed in academia and not widely used). During the third generation the cost of ALU's and control logic decreased significantly. Although parallelism through replication was now feasible, sequential languages such as Fortran, Algol and Cobol had become established. Refinements in the design and fabrication of Von Neumann style architectures allowed the trend of an order of magnitude increase in the speed of arithmetic every five years to continue.

4.1.2 Disadvantages of the Von Neumann Model

Since only one ALU exists in the model there is consequently one path (bi-directional) between the ALU and memory. A large program will typically require many operations on the data in the memory. This leads to the so-called Von Neumann bottleneck: many data items compete for the single pathway to the arithmetic unit. The amount of data waiting for entry to the ALU can be reduced by increasing the throughput of the ALU. This approach has been taken in order to reduce the Von Neumann bottleneck in current machines. The bottleneck problem, however, is inherent in serial machines and will continue to arise no matter how fast single ALU's can be made to function. This is because as new, faster processors are built, users' appetites for processing power will grow correspondingly. Also, there is an upper bound on the speed of processing that a single ALU can achieve, since this is imposed by physical constants such as the speed of light. In the next generation of Cray supercomputers the maximum wire length will be one foot, since this is the maximum distance an electromagnetic wave may travel in a single clock cycle.

A more abstract limitation of the Von Neumann model is the serial mode of problem solving it imposes. Any problem must be reduced to a series of steps to be executed sequentially [70]. Therefore, problems which have an efficient parallel solution will generate a large number of serial instructions. Perhaps a greater problem is the implementation of applications which are inherently parallel. Many problems arising in the real world are highly parallel and also non-deterministic. Consider the problem of sampling data from two sources producing signals together. The parallel solution is simply two processors sampling data from their own source. A serial implementation is much more complex. A single processor must divide its time between the sources. The non-deterministic nature of the sources means that

47

an extremely fast serial processor will be required to ensure data is not lost. Such scenarios arise in the control of nuclear power stations and weapon guidance systems. In these situations the use of replicated processors is attractive; as well as simplifying implementation, replicated resources will allow a system to maintain a processing capability as processors fail [73].

A theoretical argument can be put forward to support alternatives to the Von Neumann approach. A theoretical equivalent of parallel computation is the non-deterministic Turing machine [61], the deterministic Turing machine being the equivalent of the Von Neumann model. In the theoretical sense, the non-deterministic Turing machine is the general model and the deterministic machine a special case. The non-generality of the Von Neumann architecture accounts for the difficulty in implementing certain problems. It would seem reasonable to adopt a general model for real machines and this supports the use of parallel computers.

New programming paradigms, such as the functional approach exemplified by LISP [130] and the declarative approach of PROLOG [33], do not impose explicit sequential executions and thus are open to parallel implementations. Also the dataflow approach being investigated at Manchester [31] is specifically designed to allow parallel execution. Many of the new paradigms have a firm mathematical basis and are amenable to formal verification. Such a mathematical approach is difficult when dealing with imperative languages.

To sum up, the Von Neumann model has the following limitations:

- Lack of throughput caused by a single path between the ALU and memory, i.e., the Von Neumann bottleneck.

- Non-generality: problems must be decomposed into a series of steps to be executed sequentially; any inherent parallelism is lost in this process.

- Performance: increased processing speeds may only be achieved through hardware refinements; such refinements are becoming less cost-effective as faster circuits become more difficult to manufacture.

- Reliability: since functional units are not replicated, a single failure will suspend processing.

4.1.3 A More General Model of Computation

The original rationale for the adoption of the Von Neumann model of computer organisation no longer applies. Logic is now cheap to manufacture and microprocessors consisting of over one million gates have been fabricated [82]. Advances in technology allow the mass production of logic devices in the same way as memory. Processing units can be integrated with memory and the Von Neumann bottleneck reduced. Whilst in the Von Neumann model most of the data in memory is waiting to be processed, the memory can now be active. Wafer Scale Integration (WSI) [73] might provide a cost effective way of providing a mixture of logic and memory. WSI is the effective use of an entire wafer of silicon in a single system. At present a single wafer comprises several devices of which only a few 'good' chips can be used. WSI seeks to utilise silicon area by 'by-passing' those parts which include anomalies. Fault-tolerant chips include redundant coding; faults are detected and masked during circuit operation. Such an approach to circuit design could yield a parallel processor, three inches square, capable of 200 Mflops using today's technology.

Parallel processors can be used to achieve the processing rates of expensive serial processors for a fraction of the cost in many cases. Many areas of science generate tasks which require large amounts of computation, for example finite element simulations, artificial intelligence and many-body problems. Such applications will always require large computing resources as increasingly accurate solutions are needed. Unless fast serial algorithms are found, parallel computers must be used.

Although increased processing power has been the main impetus behind the development of parallel processors, there are other advantages in such an approach. The presence of many processors in a system should ensure that if any processor should fail, then a service can still be provided by redistributing the load among remaining processors. Also, a many-processor system should be easy to expand since extra processing capability could be gained by adding extra processors to the system.

To summarise, parallel processors promise to offer the following advantages over the Von Neumann approach:

- High performance for low cost.

- Fault tolerance.

- Extensibility.

- A more general model.

Realising the above is not trivial and will require new hardware and software methodologies. The Japanese Fifth Generation project [49], funded by government and industry, aims to provide a computer by 1992 capable of operating at 0.1-1 GLIPS (Logical Inferences Per Second; a logical inference is considerably more complex then a single instruction). This represents an increase of about four orders of magnitude in processing speed over today's computers. Such an improvement in speed will only be achieved by the use of highly parallel computers. Parallel computation has also been investigated in a decentralised way in Britain as part of the Alvey project, and also in Europe as part of the European Strategic Research Program in IT (ESPRIT).

While the parallel approach to computation represents a general framework for problem solving, this generality applies to the design of hardware platforms and software tools also. There already exist many varied parallel computers and software approaches to realising parallelism. The next section discusses how parallelism may be incorporated into the design of an architecture, how processors may be classified, and examples of the types of parallel processors currently available.

4.2 Achieving Parallelism in Hardware

Two techniques may be employed to introduce parallelism into the design of a processor: pipelining and replication.

4.2.1 Pipelining

To effect pipelining, a task must be split into a number of discrete subtasks which follow each other in time. Stages corresponding to the solution of each subtask are linked together into a linear chain or pipeline. Data entered at one end of the pipeline passes through each stage and, in doing so, is transformed to provide the required result at the output of the pipeline. At any point during the pipelining process the data associated with a single task resides in a single stage of the pipeline. Parallelism is achieved by filling the pipeline with data from many tasks, each task occupying a different pipeline stage. This type of parallelism is sometimes referred to as temporal parallelism.

The most useful analogy to use when describing pipelining is that of an industrial production line. Here, the parts of the production line correspond to pipeline stages, and the sum of these, the production line, corresponds to the pipeline. This type of parallelism is attractive since there are none of the synchronisation problems associated with the replicated systems that are discussed below. Furthermore, a pipeline reflects the natural data flow of the operation being executed. Such an approach is useful for commonly used complex operations, such as floating point arithmetic, and forms the basis for many of today's pipelined vector supercomputers such as the Cray [1]. However, hardware parallelism is often required at higher levels and, although it is possible to represent complete programs as pipelines, this requires complex pipes and is application specific. The amount of parallelism available through pipelining is limited by the number of subtasks into which an application can be partitioned. A consequence of this is that only complex tasks will yield a high degree of parallelism. As a result pipelining, although useful in some circumstances, will not provide unlimited processing power as technology develops.

51

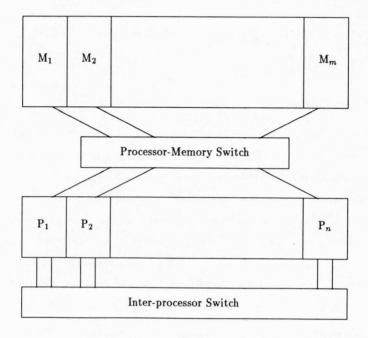

Figure 4.1: Schematic of a generic replicated system.

4.2.2 Spatial Replication

Many operations are simple yet contain a large amount of inherent parallelism. Such parallelism can be exploited by spatially replicated parallel computers. A spatially replicated system contains the following features:

- A number of processors capable of processing data in unison.

- A parallel memory from which the processors may access words together.

- A mechanism which allows processors to communicate.

Figure 4.1 [73] shows a schematic diagram of a generic replicated system. The processor-memory switch allows processors to select memories. Similarly, the inter-processor switch provides a communication path between processors. Hockney and Jessope [73] identify four parameters which serve to quantify the differences between the many parallel machines already implemented and described in the literature:

- The number of processors m, and the power of the individual processor; the product of the two determines the power of the overall system.

- The complexity of the switching network, which will determine the flexibility of the system and hence whether the power obtained by replication can be utilised by a large class of problems.

- Distribution of control to the system: whether control within the system is centralised or distributed within the system in each processor.

- Flow of control within the system: control may be derived through the execution of a predefined instruction sequence. Alternatively, control may be derived through a declarative [33] or dataflow approach [14,31,39].

Distribution of Processing Power

Two main factors affect the distribution of processing power within a multi-processor system. The first is the amount of parallelism in the problem. Clearly, a problem which is inherently serial will not execute efficiently on a parallel computer. In general, two main classes of problems are encountered:

- Highly parallel problems which can execute on large processor networks efficiently.

- Problems with a small degree of parallelism which will execute efficiently on small networks of powerful processors.

Machines which execute problems from the first class efficiently will not be efficient in general and are therefore application specific. There are enough important problems in this class, however, to justify the construction of highly parallel machines. The

DAP and the Connection Machine are examples of parallel processors with such architectures.

The second factor influencing the distribution of processing power is the ratio of cost to performance, or the efficiency, of the chosen architecture. In VLSI it is cheap to replicate simple logic. At the simplest level, processors are bit-serial and many of these work in parallel on a bit-slice of data. It can be shown that, for a given number of gates, this approach yields the maximum computational power for many simple operations. Also, the data can be stored in the exact precision required by the problem. This flexibility ensures that efficient use is made of the hardware. Although the bit-slice method leads to high efficiency for some simple operations, more complex tasks may require more complicated hardware for efficient execution.

At present most parallel processors sacrifice efficiency for speed, since chip manufacture is cheap. Such processors are complex, and relatively few are connected in networks. Parallel computers with these architectures are applicable to many problems with varying degrees of parallelism and are suitable for the execution of problems from both classes. Recently there has been a trend towards highly parallel processors because of their high performance on specific problems. As the cost of chip manufacture falls and VLSI technology improves, this trend will continue.

Switching Networks

The general problem is to establish connections between a set of inputs and a set of outputs. There are two approaches which can be taken. Direct links between input/output pairs can be made where necessary, the result being a fixed connection network. The distance between input/output pairs grows linearly as components are added to the network, but the number of links in the network follows a square law. For all but the smallest networks the complexity of the resulting network is prohibitive. This is especially true when VLSI is to be used to realise networks since only two spatial dimensions are available. Therefore, if replication is to be used on a large scale, connections must be established in a programmable way.

54

Processor/Memory Networks

In spatially replicated parallel processors two sorts of mapping between processors and memory are commonly used:

- Identity mapping: each processor has access to only a single memory. This memory is usually connected locally or is integrated with the processor itself.

- All processors may access data from any memory.

It is usual for SIMD type processors to utilise only the first mapping. MIMD computers use both types of mapping and the choice of mapping is quite important in defining the usefulness for certain applications. In general the choice of mapping divides MIMD computers into two classes, loosely and tightly coupled systems.

Loosely Coupled Systems

In loosely coupled systems, each processor has its own local memory and may not access the memories of other processors. Communication between processors is achieved by sending messages via the processor switching network. Since the cost of communication is high relative to processing, this configuration is chosen for problems where the processing to communication ratio is high.

At present the cost of such communication has led to network topologies designed to optimise performance on specific problems. As technology improves it is likely that general purpose networks with fixed topologies will be feasible. Communications costs will be reduced by the implementation of switching networks in silicon. May [91] predicts that the next generation of transputers will incorporate special purpose hardware for the routing of messages providing network transparent communication. This is already present on the Connection Machine [72,128].

Tightly Coupled Systems

In a tightly coupled system, all processors share a single global address space. Each processor may access any part of the memory, and communication takes place through references to the shared memory. There are several ways to implement a processor/memory network in such systems. These include a common bus to global memory, a crossbar switch and a packet switched network.

The cost of access to memory in a shared memory system is low, which makes such systems suitable for applications where a high degree of inter-process communication is required. As processors are added to the system, the likelihood of two processors simultaneously addressing the same location of memory increases. Memory contention decreases the effective bandwidth of communication and also requires extra resources to intervene when contention occurs; this factor places a limitation on the size of a shared memory computer. At present, the largest shared memory systems contain about 512 processors and use a packet switched network. The number of processors in a system using a crossbar switch is limited by the cost of the switch; however, this factor is not dominant in small systems. Systems which use a bus, such as the ELXSI 6400 [121] and the Sequent Balance 8000 [107], do not use many processors since the bus connecting processors and memory can soon become saturated.

As has been explained, shared memory systems cannot contain large numbers of processors using today's technology. A consequence of this is that individual processors will be relatively powerful. An extreme example of this is the Cray Y-MP [73], which is one of the most powerful supercomputers at present [85] and which contains up to four processors. In the future, loosely-coupled, message-passing parallel processors are likely to become dominant unless the problem of low bandwidth due to contention can be resolved.

The Inmos Transputer [3] represents the current state of the art in loosely coupled parallel processors. Arbitrary switching networks can be built by connecting the four on-chip links of a single transputer to other transputers. There are plans for more than four links [91], which will improve communication bandwidth and allow the construction of networks of greater complexity. Other loosely coupled parallel processors available at present tend to have fixed connection networks, examples of which are Caltec's Cosmic Cube [117] and New York University's Ultra Computer [116]. More recently, parallel processors have been constructed which allow switching networks to be changed in software [70].

Form and Distribution of Control

In general there are two control strategies which may be adopted. In the first, the flow of control is stated explicitly by the programmer before execution. In the second, the flow of control is derived from a specification of the problem/solution at execution time.

Control Flow

Most computers to date have used the control flow approach. Instructions are fetched from a memory, decoded by a control unit, and then executed on data from the memory. An algorithm which solves the specified problem is produced by the programmer. The algorithm is then decomposed into stages which can be mapped to the executable instructions of the target computer. An intermediate high level language is normally used, in preference to machine code instructions, in order to ease the implementation.

Distribution of Control

In generic, spatially-replicated systems adopting a control flow strategy, there are two alternatives for the distribution of control. Firstly, there may be one control unit decoding a single sequence of instructions and distributing the results to many processors. More generally, each processor may have a control unit processing its own sequence of instructions autonomously. The distribution of control and data has formed the basis for a simple taxonomy of control flow computers by Flynn [54].

Flynn's Taxonomy

In this classification scheme, the flow of instructions is related to the flow of data. A stream is a sequence of items (instructions or data) which are executed/operated on by a processor. A stream may be single or multiple, yielding four categories:

- SISD Single Instruction stream, Single Data stream: single instructions operate on single data items; the Von Neumann model is in this class.

- SIMD Single Instruction stream, Multiple Data stream: single instructions operate on many data items at once; machines such as the DAP [111] and the Connection Machine [72] belong to this class.

- MISD Multiple Instruction stream, Single Data stream: this class seems to be empty since it is hard to attach any meaning to several operations being performed simultaneously on a single data item. If the data in a pipeline is regarded as a single item, then a pipelined computer can be considered a member of this class.

- MIMD Multiple Instruction stream, Multiple Data stream: multiple processing units execute their own sequence of instructions on their own data in an asynchronous way.

This taxonomy divides parallel computers into two classes according to whether control is local or centralised. The scheme is too general to be used to characterise parallel computer architectures in any detail since all parallel machines are placed in the SIMD class except for those with multiple control units: for this reason the taxonomy is generally used as a shorthand.

SIMD Machines

In this class we have pipelined machines such as the Cray 1 and Cray 2 [1] and also processor arrays such as the AMT DAP, BSP [17,76], ILLIAC IV [45] and the Massively Parallel Processor [106]. The availability of only a single instruction stream might seem a limitation when extracting parallelism from a problem. Many problems, however, require the application of the same instruction to many items of data. Homogeneous problems of this type are quite common and ideal for implementation on a SIMD type machine. This type of parallelism is referred to as data or structure parallelism.

May [91] states that the parallelism available in processor arrays will be limited by the overhead of passing the instruction data from the control unit to the processing elements. This is a valid criticism at present but when VLSI is utilised and whole processor arrays are implemented on the same piece of silicon, the communication cost will be small. An example of such a processor is the Blitzen [27] chip on which 128 processing elements together with local memory are fabricated on one piece of silicon, comprising over one million transistors. Using such technology a 16k PE SIMD processor could be built from just 128 chips.

MIMD Machines

MIMD machines can be used effectively on a wider class of problems than can SIMD processors. Since a control unit is present in each processor, MIMD processing elements are usually more powerful than their SIMD counterparts. If this was not the case, then the overhead in fetching and decoding an instruction might be greater than in performing the actual operation.

Although MIMD processors provide greater flexibility than SIMD machines, their asynchronous nature introduces new problems for the system designer. Gajski and Peir [57] identify three basic problems associated with MIMD computing:

- The partitioning problem: partition a problem into tasks.

- The scheduling problem: assign each task to one or more processors for execution.

- The synchronisation problem: assure an order of execution that leads to correct results.

The first two difficulties are common to SIMD and MIMD processors. Synchronisation is performed by hardware in the SIMD case, while the programmer must ensure that processes are synchronised on a MIMD processor. If processes are not synchronised correctly, deadlock may result. Deadlock avoidance is difficult because of the inherent non-determinism of MIMD processors. Hockney and Jessope give an example of non-determinism:

- If there are n instruction streams and no synchronisation present, after a single instruction has been executed there are n possible states that the system might be in. After the next instruction the number of possible states is n^2, etc.

This exponential growth in possible states makes MIMD systems unpredictable. It is quite possible for the programmer to overlook certain states occurring, the result of which might be incorrect execution or deadlock; moreover, this could occur at any time. It is not uncommon for a program bug to disappear when code is added to trace the system's behaviour: this is a manifestation of non-determinism.

An interesting example of a MIMD processor is the Cosmic Cube developed at Caltech [117], in which standard Intel i860 processors [82] are connected in a hypercube network. More recently processors have been specifically designed for the implementation of MIMD parallel processors. The transputer [3] is described as a computer on a chip. Fast local memory, a floating-point processor and four data links reside on the same silicon as the main processor, and all may operate simultaneously. MIMD machines can be constructed by connecting transputers, which form the building blocks for the parallel architecture.

The control flow approach can lead to extremely efficient programs on both SIMD and MIMD machines for many types of problem. However, the programmer is forced to state explicitly the sequence of instructions that will lead to a correct solution. This task is not difficult when the problem is homogeneous and the target machine is of SIMD type. However, problems often contain a mixture of data structures which will not easily map onto such a machine. In such cases MIMD machines can be used; but the programmer is then faced with non-determinism and its side effects.

Such difficulties have led to new programming paradigms and architectures being investigated. Functional programming and the dataflow approach are alternatives to the control flow model.

Dataflow

In the dataflow method [39], a statement is regarded as being ready for execution if the data that it requires is available. The order of execution of statements therefore depends on the flow of data in the system and not the textual order of the instructions. If data is ready for more than one statement, they may be executed in parallel. Dataflow languages provide a syntax which allows the programmer to describe a problem's solution while making data dependencies explicit.

When executing a dataflow program it is first necessary to isolate data dependencies in order to determine the correct order of execution and also extract any parallelism. A graph is constructed where nodes represent statements, or at a lower level operators. Arcs connecting nodes pass data between nodes. A node is said to fire when it executes its respective statement. A node may not fire until data is present at all of its input arcs. The same techniques used in optimising compilers for parallel computers are used to analyse data dependencies and to construct the graph [118]. Consider the following sequential code:

1. A=A+C;

2. D=A*C;

3. E=A/B;

4. F=A+C;

5. G=(D+E)*F.

The graph shown in Figure 4.2 may be constructed where node numbers identify statements. In the dataflow model, once statement 1 has been executed statements

61

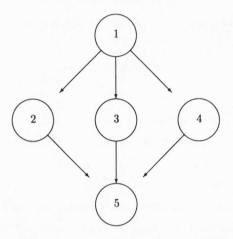

Figure 4.2: Graph showing data dependencies.

2, 3 and 4 may be executed together. The sequential model calculates the value of G in five steps, compared with three for the dataflow model.

The level of parallelism produced by the dataflow approach is too fine for conventional control flow parallel computers to execute efficiently [118] (see Section 4.5.2); therefore, several specialised architectures have been proposed [118]. One such architecture has been implemented by a group at Manchester University. The Manchester Dataflow machine [31] has been operational since 1981 and performance rates of 1-2 MIPS have been measured on reasonably large problems [103]. Other important groups working on dataflow are at MIT.

The Functional Approach

The functional approach to the generation of control has been used for many years. The functional language LISP was one of the first high level programming languages and is still widely used, mainly in the field of artificial intelligence. Functional languages are based on the mathematics of the Lambda Calculus (a mathematical notation for describing algorithms). Most functional language implementations have features which ease programming but also render them not as mathematically 'pure' as the Lambda Calculus. Even so, functional implementations of systems are still more amenable to formal verification than conventional languages. Also, many specification languages have a strong functional basis, for example Z [67]. Such a specification will lead to a simple functional implementation in most cases. The theoretically attractive features of functional languages have led to their use outside the AI domain.

A program in a functional language may be regarded as a call to a single function. The arguments of this function might be calls to other lower level functions, and so on. A tree can be constructed representing a program's component function calls; nodes in the tree represent function calls and branches denote the evaluation of function arguments. The leaves of the tree correspond to arguments which may be evaluated directly, i.e., constants, and thus require no further evaluation. The process of generating branches of the tree is known as reduction. Consider the LISP program:

$$(\text{times}(\text{plus } 3(\text{divide } 4 \text{ } 2))7),$$

which evaluates the expression $((3+(4/2))*7)$. The program may be represented by the tree shown in Figure 4.3. In pure functional languages branches are completely independent; thus the reduction of different branches may take place independently. Each subtree generated by a node may be evaluated using parallel hardware and this has been the basis for parallel implementations of functional languages. Currently, there is much interest in this area [37,104]. However, this field is not so far advanced as that involved with dataflow techniques [73].

When dataflow and functional programs are executed on serial machines, many instructions are executed for every useful instruction; this leads to poor efficiency. This is because Von Neumann architectures are optimised for the control flow model. In the future there will be a shift towards the use of special purpose architectures for non-control flow implementations. When such systems offer similar performance levels to conventional computers, the higher level of abstraction offered will make

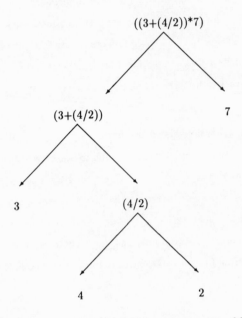

Figure 4.3: Tree showing order of evaluation in a functional language.

their use attractive.

4.3 Other Taxonomies

Flynn's taxonomy has already been covered in order to provide a concise way of describing different control strategies. There are several other commonly cited classification schemes, of which two will be described.

4.3.1 Shore's Taxonomy

Shore's taxonomy classifies architectures according to how they are organised from four basic parts: the processor unit, the control unit, data memory and instruction memory. Six basic architectures are identified, four of which subdivide the SIMD category of Flynn, although the class of pipelined vector computers is not satisfactorily covered [73]. The taxonomy serves as a useful, though far from comprehensive, extension of Flynn.

4.3.2 ASN (Algebraic Structure Notation)

Hockney and Jessope [73] have proposed an algebraic notation which describes the number of instruction, execution and memory units, and their inter-connection and control. The notation regards a computer as comprising a number of interconnected functional units operating under the control of instruction units. In all, 20 rules are used to describe the symbols representing units, how to connect them into groups, connections between units, control of units and comments. Instead of a finite set of architectures, the taxonomy allows architectures of arbitrary detail to be described, and the set/language of acceptable descriptions can be described in Bachaus-Naur form. The large number of rules enables detailed descriptions of processors. However, this will not encourage usage, because of the resultant complexity of coding/decoding descriptions in the taxonomy. For example, the ASN description of a DAP 610 is:

$$C(DAP\ 610) = C[64^2\overline{P}]_l^{1-nn}; P = B_1 \text{-} M_{16k*1}$$

Although Flynn's taxonomy groups all parallel computers into just two groups, it continues to be used widely. In the future it is likely that a more rigorous taxonomy will be required, in which case a scheme similar to the one just described may become popular.

4.4 Hardware Performance Measurements

Hockney and Jessope [73] have proposed two parameters, R_∞ and $n_{\frac{1}{2}}$, for measuring the peak hardware performance of processors when executing arithmetic operations on a vector. The first parameter, R_∞, is the maximum or asymptotic performance, and is defined as the maximum rate of computation in floating point operations per second. For a generic parallel processor with an infinite number of processing elements, R_∞ is achieved for vectors of infinite length, and hence the ∞ subscript. This parameter is often used when comparing the performance of serial machines. In this context, it has been criticised since high arithmetic performance does not guarantee fast processing rates for general problems. Also, no account is taken of the precision of arithmetic, and this will vary for machines of different word lengths.

The second parameter $n_{\frac{1}{2}}$ is the half performance length. It is defined as the length of the vector required to achieve a processing rate of $\frac{R_\infty}{2}$; $n_{\frac{1}{2}}$ measures the amount of parallelism in the architecture. For a serial processor $n_{\frac{1}{2}} = 0$ and for an infinite array of processors $n_{\frac{1}{2}} = \infty$. Here, the type/precision of floating point operations is irrelevant and this is perhaps the more useful of the two measures.

4.5 Parallel Algorithms

In order to exploit the parallelism of an architecture, a parallel algorithm must be developed. Quinn [107] recognises three ways in which a parallel algorithm may be designed:

- Detect parallelism in an existing algorithm.

- Adopt a known parallel algorithm which solves a similar problem.

- Invent a new parallel algorithm.

The first route may not be possible since the serial algorithm may have a large sequential component. Quinn [107] points out that blindly transforming a serial algorithm into a parallel one may lead to large inefficiency. Adapting a known parallel algorithm which solves a similar problem is the most attractive of the three possibilities. As the parallel software base grows the number of programming paradigms will increase. Consider the field of AI, where many problems can be formulated as state-space searches: once one parallel search algorithm has been developed it may be adapted for individual cases. Finally, it may be necessary to invent a new algorithm: the first two methods may not be possible or may lead to inefficient implementations; also, the target machine may demand special consideration in order to optimise performance.

At present, many different parallel architectures exist. The performance of an algorithm can vary considerably on different hardware platforms. It is therefore quite important to match algorithm to architecture. There are several reviews of parallel algorithms [69,115]. Kung identifies three important attributes of an algorithm:

- concurrency control;

- granularity;

- communication geometry.

4.5.1 Concurrency Control

This is the control which ensures correct execution of the program. The concurrency control enforces the synchronisation between processes which will lead to successful completion. The control may be centralised, in which case synchronised execution results, or distributed, which leads to asynchronous execution.

4.5.2 Granularity

The granularity of a parallel algorithm is the ratio of processing to communication and can range from fine (frequent synchronisation) to coarse (computation dominates).

4.5.3 Communication Geometry

This is the geometry of a network representing all inter-process communication.

The attributes listed in Section 4.5 can be shown to correspond with hardware. The most obvious connection is for communication geometry which matches with the inter-processor switching network. Often there will not be a direct mapping between

67

the two since the communication geometry will be too complex to implement in hardware.

The granularity of a processor is usually defined as the amount of computation performed by a processing element before synchronisation, or the ratio of processing to communication. In SIMD processors, synchronisation takes place after every program instruction, and the granularity is thus fine. In MIMD computers the grain size is generally larger, since processing elements can work asynchronously for arbitrary lengths of time. Synchronisation in MIMD computers is expensive and it is therefore important to keep processors busy; a fine-grained algorithm is therefore unsuitable for implementation on a MIMD processor. Conversely, a coarse-grained algorithm will perform poorly on a SIMD machine because a large amount of time will be spent needlessly achieving synchronisation. Concurrency control corresponds to the distribution of control. An algorithm in which concurrency control is centralised will be suitable for implementation on a SIMD type machine, whereas a MIMD implementation would be inefficient because of wasted control resources. Algorithms which perform the same operation on large data sets are members of this class. The data can be partitioned and distributed among processors; each processing element will execute a global instruction on local data: this is sometimes referred to as geometric or data parallelism [70]. Algorithms which utilise decentralised, asynchronous control will require the multiple control units of a MIMD processor. An example of such an algorithm is one in which a central resource generates tasks and collects results after execution, and in which the tasks are independent and do not require synchronisation. Algorithms of this type may be implemented on a processor farm [70]. A central processor produces and distributes tasks to 'worker' processors which are not busy; after execution of a task, a worker processor sends the results back to the master processor.

4.5.4 Measuring Algorithm Performance

Speedup

R_∞ is a useful measure of optimum hardware performance, but in practice it will not be achieved. This is because the rate of useful processing is a factor of both the algorithm and the hardware. This is implicit in the concept of 'speedup' which is defined as the ratio of performance of a parallel algorithm running on n processors to the same algorithm executing on a single processor:

$$S(n) = \frac{t_1}{t_n} \tag{4.1}$$

t_1 = time for execution on uni-processor,

t_n = time for execution on n processors.

This measure can be misleading since most parallel algorithms include extra instructions to accommodate parallelism. In these cases the value for t_1 will be artificially high and the overhead of extra instructions will be hidden, with the result that speedup will be exaggerated [107]. Also, on SIMD processors, where individual processing elements have little power, the value t_1 will be very high and will produce meaningless values for speedup [101,102].

Alternatively t_1 can be defined as the time taken by the fastest serial machine to execute the best serial algorithm. In practice a well known serial machine is used to measure t_1 since the most powerful serial computers are often not available for researchers. This provides a more meaningful indication of whether a parallel implementation will be cost-effective.

In this thesis the latter definition of speedup will be used to measure the performance of parallel algorithms. The serial processor used to compare performance with the DAP 610 will be the University of Sheffield IBM 3083 BX mainframe computer.

Linear Speedup

A parallel algorithm is said to exhibit linear speedup if the speedup achieved with n processors is $O(n)$ or

$$S(n) \in O(n).$$

Linear speedup is hard to achieve because of overheads such as contention for shared resources, communication and the difficulty in structuring software so that arbitrary numbers of processors are kept busy. Minksy [93] noted that typically speedup has the form $S(n) = \log(n)$.

Superlinear Speedup

Superlinear speedup is achieved if

$$S(n) > n.$$

One argument says that this is impossible since a serial machine can simulate a parallel machine through time-slicing [44]. This presupposes, however, that the serial algorithm is chosen after the particular problem instance is decided. If serial and parallel algorithms are chosen beforehand, extra processors may enable the parallel algorithm to exhibit superlinear speedup by finding solutions very quickly in some cases.

Efficiency

A related measure to speedup is efficiency. Efficiency is defined as the average utilisation of processors [42]:

$$E(n) = \frac{S(n)}{n}. \tag{4.2}$$

Efficiency is quite important on a MIMD machine because processors are usually costly. Several studies have been carried out into the relationships between speedup, efficiency and software parallelism [48,68]. Eager *et al.* [42] consider the inherent tradeoff between speedup and efficiency in a software system. In particular,

70

as maximum speedup is approached, efficiency may degrade linearly to arbitrarily low values. In such cases maximum speedup might be costly to attain. Efficiency is less important in SIMD processors because the processing elements are simple and large numbers can be deployed.

4.5.5 Amdahl's Law

Amdahl's law is a good argument against the usefulness of large scale parallelism. Often an algorithm will have an inherently serial component and this will limit speedup. Amdahl's inequality relates the maximum speedup achievable to the fraction of the algorithm which is inherently serial:

$$S(n) \leq \frac{1}{s + \frac{1-s}{n}} \tag{4.3}$$

$n = $ no of processors,

$s = $ fraction of code which is serial.

As $n \to \infty$, the maximum speedup $\to \frac{1}{s}$. Thus, if five percent of the operations are sequential the maximum speedup will be 20. However, in some cases s can be made quite small through techniques which reduce non-overlapped communication, load imbalance and sequential operation dependency [63]. Also, many important algorithms have a near negligible sequential part.

Gustafson *et al.* [63] suggest that it may, in fact, be easier to achieve a good speedup than one might infer from Amdahl's law. This is made possible by scaling the problem size with the number of processors. The inverse of Amdahl's paradigm is considered: that is, how long a given parallel algorithm will take to execute on a single processor. This leads to an alternative to Amdahl's law called scaled speedup:

$$\text{scaled speedup} = \frac{s\prime + p\prime n}{s\prime + p\prime}$$

$$= n + (1 - n)s\prime \tag{4.4}$$

71

sℓ = serial fraction,

pℓ = parallel fraction,

n = no of processors.

Two assumptions are implicit in Equation 4.4. Firstly, it is assumed that only the parallel part of a program scales with problem size. Secondly, it is assumed that the amount of work that can be done in parallel varies linearly with the number of processors. Both of these assumptions are valid for certain types of problem. However, it is not fair, in general, to assume that an application's serial content remains fixed for different problem sizes. Therefore, although scaled speedup may provide an insight into parallel performance for some applications, it is not valid in the general case for which Amdahl's law still applies.

In this chapter we have discussed the present state of parallel processing systems. In the next chapter, we discuss the particular parallel processor used in the experiments described in this book.

Chapter 5

DAP Hardware and Software

5.1 Introduction

The parallel processor used for this work was the AMT DAP 610, hosted by a VAX 8350 minicomputer, at Queen Mary and Westfield College, London. The machine was accessed from Sheffield by using JANET (Joint Academic Network).

No generally accepted program development methodology for parallel computers has been established; one technique is to extend existing languages to accommodate parallelism. This is particularly applicable in the case of SIMD computers which closely identify with the Von Neumann model [108] since both employ a single control unit. Such an approach is used when programming the DAP. An extended version of Fortran 77, known as DAP Fortran or Fortran Plus [2], is used,which incorporates parallel array handling facilities.

In order to utilise the massive parallelism of the DAP it is important to understand the differences between the DAP and conventional computers. This chapter describes in detail the architecture of the DAP, and this is followed by a description of Fortran Plus and how its extensions reflect the structure of the DAP.

73

5.2 DAP Architecture

The DAP shares the three main components of the Von Neumann model:

- Control Unit.

- Memory.

- Arithmetic/Logic Unit (ALU).

Parallelism through spacial replication is achieved within the DAP ALU and memory as shown in Figure 5.1. The Master Control Unit (MCU) of the DAP fetches

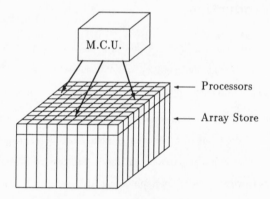

Figure 5.1: Schematic of the DAP.

instructions from the code memory. Instructions are either executed entirely within the MCU or within the processor array. Array instructions must be broadcast to each and every processing element (PE); such instructions are then executed by the PE on local data.

5.2.1 The Master Control Unit

The MCU is a 32-bit processor, which in the latest generation of DAPs has a cycle speed of 10 MHz; most instructions finish in a single cycle. The MCU has many of the features of a conventional processor including registers, an ALU for scalar arithmetic, an instruction counter and conditional branching. Most MCU instructions deal with data stored in the array and are executed by the PE array. A smaller subset of instructions is confined to use within the MCU; these instructions operate on MCU registers and alter control within the program. Other functions of the MCU are:

- Transmission of data between the array memory and MCU registers.

- The support of data transfer between the DAP and the host filestore or attached peripheral devices.

The MCU will execute certain instructions more efficiently than the processor array. Examples are scalar arithmetic and branching since the MCU is optimised for these purposes. A further characteristic of the MCU is the ability to execute loop instructions efficiently since this facility is supported by specialised hardware loop primitives.

5.2.2 The Processor Array

The processor array consists of many PE's joined together to perform instructions broadcast by the MCU. The processor array is the heart of the DAP's parallel processing capability: each PE is very simple and it is the combination of many such basic units which allows the DAP to achieve high processing rates.

The DAP Processing Element

Figure 5.2 shows a simplified diagram of a DAP PE. The DAP PE is a 1-bit processor, which means that all logic, arithmetic and transfers to/from memory are made 1-bit at a time. Such simplicity permits a PE to be constructed from relatively few gates occupying a small area of silicon; silicon is ideal for replication of such elementary logic. Operations on data lengths greater than a single bit must be controlled

75

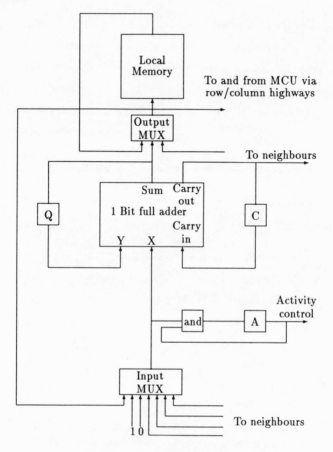

Figure 5.2: The DAP Processing Element.

by software and the speed of execution is directly related to the data length. Each PE has three 1-bit registers:

- Q - accumulator.

- C - carry.

- A - activity control.

The arithmetic and control features of a DAP PE are relatively simple and can be described in the single paragraph which follows. The activity register determines

76

whether or not the current instruction is to be executed. This allows a subset of PE's to be selected, based on local conditions. The only arithmetic the PE can perform is a 1-bit addition (an extreme example of a RISC architecture), where the inputs are Q, C and a multiplexed input from any of the following: Q, A, carry from a neighbouring PE, MCU, local memory. The A register also derives its input from the multiplexer and may be written to directly, if ANDed with its existing contents; the multiplexer may also be inverted. The output from a PE may be written to memory, conditional on the A register if necessary. A further register, D, is used by the fast interface unit for asynchronous movement of data during processing. There is no hardware within the PE to generate memory addresses and this is done centrally within the MCU. A corollary of this is that the PE's cannot address local memory independently; instead, PE's all access the same local address determined by the MCU. The advantage of such an approach is that the PE architecture is greatly simplified at the expense of flexibility. The minimalist approach to PE design described permits the use of massive parallelism at a low cost. The Connection Machine [71] processing element may access memory independently, but this is at the expense of more complex hardware.

Inter-connection of the Array

Processing elements are connected in a square 2-D array, with each PE connected to its four nearest neighbours as shown in Figure 5.3. PE's at the edges of the array are connected to those on the opposite boundary in a wrap-around fashion resulting in a torus geometry. PE's in each row and column are connected by data highways which allow data to be broadcast/received, from the MCU/all PE's, simultaneously. Both local and row/column highways are bit-serial.

The size of the DAP array is determined by the edge size ES, which is the number of PE's along one side of the array. In the DAP 510, $ES=2^5=32$, yielding 32*32=1024 PE's; the last two digits of the model number indicate the clock speed. Currently the DAP is available in two sizes: the larger in terms of edge size is the

77

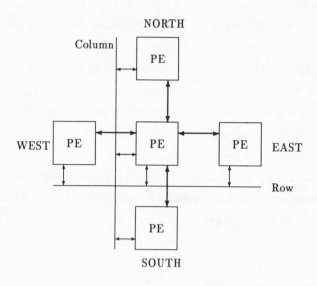

Figure 5.3: Inter-connection of the Array.

DAP 610 which has 4096 PE's. All PE's can simultaneously transfer 1 bit to/from memory and also perform a logical operation on the data in a single cycle, this resulting in a processing rate of 10^4 MIPS (million instructions per second) for a DAP 510 and $4*10^4$ MIPS for a DAP 610. Nearest neighbour connection is useful in many data parallel applications which require successive updating of local memory based on neighbour information. The row/column highways provide an efficient way of sending the same information from the MCU to every processor and for collecting data from the processor array.

The Array Store

Each PE has its own local memory, the size of which can vary between 32 Kbits and 1 Mbit depending on the particular DAP that is being used. The collective memory of all PE's is referred to as the array store. The same address in every local PE

78

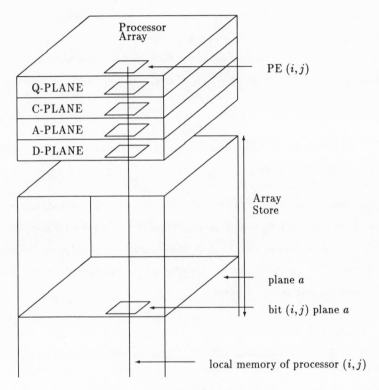

Figure 5.4: Array Store.

memory is collectively referred to as a storage plane and, similarly, the set of each of the registers Q, C, A and D is known as a register plane. The array store may be regarded as a 3-D structure, as shown in Figure 5.4, consisting of a number of horizontal planes of size $ES * ES$ arranged vertically.

Since the DAP is a bit-serial processor it is not committed to any particular representation of data and all data may be treated as a sequence of bits. Data representation is determined by software, which results in flexibility and efficiency. Two frequently used representations are referred to as horizontal and vertical modes:

- Horizontal mode - the data item is mapped within a single storage plane.

- Vertical mode - the data item is mapped into the same position within successive storage planes.

In horizontal mode, a plane can be divided in 32-bit words, so that each row will correspond to a single word on a DAP 510 and to two words on a DAP 610: this mode is used in Fortran Plus to store scalar values. Vertical mode is suited to massively parallel operations where each PE contains a separate data item. Fortran Plus supports both modes of storage.

5.2.3 DAP I/O

Figure 5.5 shows the hardware stages involved in linking the DAP to a host computer. The DAP is connected to a host computer via the Host Connection Unit (HCU), which serves as a communications gateway between the two. The HCU is driven by a Motorola 68020 and the kernel is stored in a 256k EPROM. The HCU includes a SCSI interface for connection to SUN-based hosts. A VME bus allows connection of standard interface cards and also connection with VAX-based hosts. The HCU is used for medium speed data transfer of up to 1 Mbyte/sec; also provided are fast data channels which can operate at up to 40 Mbytes/sec full duplex. Data transfer at this rate requires only 0.8% of a DAP 610's processing cycles and 3% of a DAP 510's.

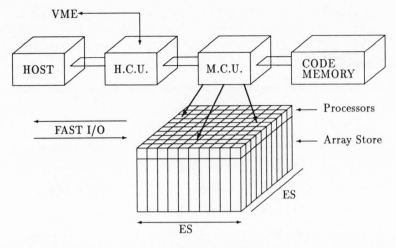

Figure 5.5: DAP I/O.

The D plane is used for fast I/O. For output, a storage plane is copied into the D register, and the D register is then shifted out to the fast data channel one row at a time from the northern edge. For input, data is shifted from the fast I/O channel onto the southern edge of the D register one row at a time and the complete D plane is then copied to array store. Shifting is completed autonomously and the MCU is only interrupted when transferring data to/from the array store, the transfer taking a single cycle. Thus fast I/O proceeds concurrently with normal processing. At present two devices are available which utilise the fast I/O facility: the fast video interface allows real-time data visualisation of large amounts of data at high resolution, while a fast disk unit allows rapid retrieval of data from backing store.

5.3   Program Development on the DAP

An application designed to run on the DAP consists of two programs:

- A host section which may be written in any language supported by the host. This program deals with all the I/O and interaction with the host operating system.

- A DAP section written in a language containing parallel extensions, the language generally being Fortran Plus. In certain circumstances the developer may choose to use APAL (Array of Processors Assembly Language).

Both programs are developed on the host computer using the program development facilities provided by the host in the usual way. The host program is used to send/receive data from the DAP section since the DAP may not, under normal circumstances, perform I/O independently of the host. The host program will then invoke the DAP code when required. The usual approach is for the DAP code to comprise the compute-intensive part of an algorithm, with the remainder running on the host.

The next section will describe the parallel extensions delivered by Fortran Plus. The array extensions provided are similar to those proposed in the new Fortran 9Y draft standard.

5.3.1 Fortran Plus

Fortran Plus contains features which allow the programmer to manipulate complete data structures as single objects. As well as the scalar data type, two further data types exist. These are:

- Vector - a one-dimensional array of size ES.

- Matrix - a two-dimensional array of size $ES * ES$.

Operations on these objects are performed in parallel on every object. Extra facilities are provided to manipulate vector and matrix objects and these may be divided into three classes:

- Parallel execution of instructions on all elements of the object with no need for loop constructs.

- Operation on selected subsets of elements within the object.

- Special functions which manipulate vectors and matrices.

Data Types and Precision

All Fortran 77 data types are permitted in Fortran Plus, but Fortran Plus allows a greater range of data lengths to be specified. Integers may be stored in anything between 8 bits and 8 bytes. Real numbers may be allocated between 2 and 8 bytes. When vector and matrix modes are used the declared data length reflects the actual amount of storage used, with logical variables occupying a single bit. Examples:

INTEGER*3   *smallint*   3 byte integer scalar

REAL*7   *bigreal*   7 byte real scalar

Vector and matrix objects are declared by leaving empty subscripts in the first dimension for a vector and the first two dimensions for a matrix. These dimensions are referred to as 'constrained' dimensions, since they are constrained to the physical dimensions of the DAP, for example:

CHARACTER *phrase*()   a vector of characters

INTEGER*4   *potential*(,)   a matrix of 4 byte integers

Arrays of such objects may be declared by adding the appropriate extra dimensions:

CHARACTER *phrases*(,5)   5 character-vectors

INTEGER*4   *potentials*(,,3,2)   a 3*2 integer array of matrices

Scalar, Vector and Matrix Modes

Variables of mixed type, precision and mode may be combined in Fortran Plus expressions. Fortran 77 rules are used for combining objects of different type. An expression involving variables of different precision will have the same precision as the variable of longest length. Thus given

INTEGER*6   *longint*

REAL*4   *x*

REAL*8   *longreal*

the expression

$(longint + x)/longreal$

will have type REAL*8.

When combining objects of different modes, variables are expanded to conform (have the same number of elements) as the largest object. Coercion is achieved by replicating the existing element(s) over the extra dimension(s). For example, if we have the declarations:

INTEGER*4 $x(,), y(,), z$

$z = 5$

$$x = \begin{vmatrix} 2 & 7 & 3 \\ 4 & 2 & 1 \\ 5 & 3 & 3 \end{vmatrix}$$

then the expression

$y = x * z$

will result in

$$y = \begin{vmatrix} 10 & 35 & 15 \\ 20 & 10 & 5 \\ 25 & 15 & 15 \end{vmatrix}$$

The following table shows the resultant mode when combining two sub-expressions; sub-expression 1 is shown on the x axis and sub-expression 2 on the y axis.

	scalar	vector	matrix
scalar	scalar	vector	matrix
vector	vector	vector	matrix
matrix	matrix	matrix	matrix

Indexing

Fortran Plus allows the selection of scalars from arrays in the same way as Fortran 77. Extra indexing mechanisms are also provided to allow the selection of Fortran Plus objects. It is also possible to choose a subset of elements from a vector or matrix for assignment. These two types of indexing are referred to as right and left hand indexing respectively.

Right Hand Indexing In Fortran 77 it is possible to select a scalar from an array of items by specifying subscripts in the appropriate dimensions. Fortran Plus also allows the selection of objects of other modes from arrays of such objects. It is also possible to select a vector from a matrix. The following table shows the permitted instances of right hand indexing. The type of the right hand variable is shown on the x axis, and the type of the left hand variable on the y axis.

	scalar array	vector	vector array	matrix	matrix array
scalar	yes	yes	yes	yes	yes
vector	no	yes	yes	yes	yes
matrix	no	no	no	yes	yes

Selection of a Vector Selection of a vector is done by including a null subscript in the required parallel dimension. For example, if we have:

INTEGER $x(,), y()$

$$x = \begin{vmatrix} 4 & 14 & 6 \\ 8 & 4 & 2 \\ 10 & 6 & 6 \end{vmatrix}$$

85

then the expression

$$y = x(3,)$$

gives

$$y = \begin{vmatrix} 10 & 6 & 6 \end{vmatrix}$$

Alternatively an index vector may be used to select the appropriate elements:

INTEGER $index()$

from each row/column:

$$index = \begin{vmatrix} 3 & 1 & 2 \end{vmatrix}$$

then

$$y = x(, index)$$

gives

$$y = \begin{vmatrix} 10 & 14 & 2 \end{vmatrix}$$

A further method involves the use of a logical index matrix which contains a single TRUE value in each row/column, the position of the TRUE indicating which element is to be selected:

LOGICAL $index(,)$

$$index = \begin{vmatrix} F & T & F \\ T & F & F \\ F & F & T \end{vmatrix}$$

$$y = x(index,)$$

gives

$$y = \begin{vmatrix} 8 & 14 & 6 \end{vmatrix}$$

This method may also be used to select scalars from vector/matrices by using a logical object of the same mode with a single TRUE element.

Left Hand Indexing The techniques used in right hand indexing may also be used on the left hand side of an assignment statement. In this case the construct specifies which elements are to be assigned to:

86

INTEGER $x(,)$, *scalar*

$x(2,3) = scalar$

Logical indexes are referred to as 'masks' when used on the left hand side of an assignment statement. In this case a mask may have more than one TRUE element. Every element of the object which corresponds to a TRUE element in the mask is updated. For example, a logical mask can be used to prevent division by zero:

REAL $x(,), y(,), z(,)$

$x(z.\text{ne}.0) = y/z$

The equivalent serial code would be:

DO 10 $i=1,64$

    DO 20 $j=1,64$

        IF $(z(i,j).\text{ne}.0)$ $x(i,j) = y(i,j)/z(i,j)$

20   CONTINUE

10   CONTINUE

Logical masking is performed very efficiently on the DAP and corresponds to setting the activity control registers within PE's. This type of indexing is used extensively in Fortran Plus programs to accommodate multiple data streams. In a serial program there is a single data stream and the control path may be changed by the use of branching constructs such as IF THEN ELSE. For a SIMD processor there may be many possibilities, one for each stream of data. Branching is not a solution since only a single control path may be followed, and in such circumstances a mask may be used to select each alternative data stream in sequence. For example:

Transform the data in x as follows:

$$x = \begin{cases} x^2 & x > 0 \\ -x & \text{otherwise} \end{cases}$$

Serial Implementation:

REAL $x(64,64)$

DO 10 $i=1,64$

87

```
      DO 20 j=1,64
          IF (x(i,j).gt.0) THEN
              x(i,j) = x(i,j) * x(i,j)
          ELSE
              x(i,j) = -x(i,j)
          ENDIF
   20    CONTINUE
   10 CONTINUE
```

Parallel Implementation:

```
      REAL x(,)
      x(x.gt.0)=x * x
      x(x.le.0)=-x
```

Long Vector Mode

It is possible to address a matrix variable as if it were a one-dimensional vector of length $ES * ES$: this is referred to as long vector addressing.

Functions and Subroutines

Fortran Plus has many of the intrinsics of Fortran 77, some of which are extended to work with Fortran Plus objects. It is also possible for the user to develop such procedures. The language also includes many functions, known as aggregate functions, which operate on vectors and matrices. These functions implement commonly required operations, and complex programs may be built from such primitives. Examples:

- MAXV - takes a vector/matrix and returns the scalar which is the largest element of the object.

- TRAN - takes a vector/matrix and returns the transpose of this matrix.

- SUM - returns the sum of the elements of a vector/matrix.

88

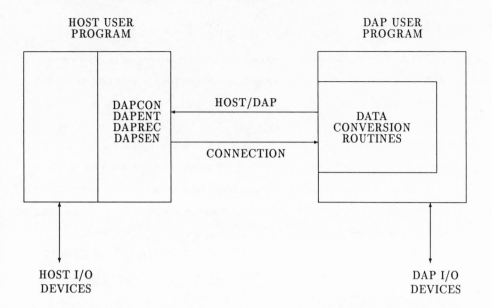

Figure 5.6: Structure of a DAP Program.

Such functions have very efficient implementations and harness the bit-level processing power of the DAP. The execution times of these functions are related to the precision used.

5.3.2 The Structure of a DAP Program

In any DAP system, a host computer provides all operating system facilities for the DAP, which is regarded as an attached processor. Although the DAP can communicate with the outside world, Fortran Plus has no I/O facilities and relies on the host for such activities. Hence a host program must exist to send data to/from the DAP. Fortran COMMON blocks are used for this purpose. A further complication is that data is represented differently in the host and DAP. Therefore after data is sent/received it must be transformed to the appropriate storage format. Often different representations are allocated differing amounts of storage. For example, on a serial machine most variables are stored as complete words whenever possible, for processing efficiency. On the DAP, however, data is usually stored in the smallest possible space, with no wastage occurring.

Any DAP application consists of two programs, one running on the host and one running on the DAP. The DAP is viewed as a resource by the host program and it must therefore request access to the DAP before DAP processing may proceed; this is accomplished by calling the function DAPCON. After access has been gained the host may send data to the DAP using the DAPSEN subroutine; to receive data the analogous subroutine DAPREC is used. DAP processing is invoked by a call to a DAP 'entry' subroutine, which executes on the DAP. The host subroutine DAPENT is used to pass a string containing the name of the DAP entry subroutine; after the RETURN statement is executed in the entry routine, control is returned to the host program. Finally, the DAP resource is released by executing a call of DAPREL. The following example shows the structure of the host and DAP programs (see also Figure 5.6):

Host Program

```
PROGRAM HOST
INTEGER data(64, 64), dapcon
COMMON/B1/data
IF (dapcon('dap_filename').eq.0) THEN
        CALL DAPSEN('B1',data,4096)
        CALL DAPENT('DAPENT')
        CALL DAPREC('B1',data,4096)
        CALL DAPREL
ELSE
        PRINT *,'failed to connect to DAP'
        STOP
ENDIF
```

DAP Program

```
ENTRY SUBROUTINE DAPENT
INTEGER data(,)
COMMON/B1/data
{convert data to DAP format}
{process data}
{convert data to host format}
RETURN
END
```

Data conversion is achieved by calling built-in conversion routines.

A more detailed description of DAP hardware and software may be found in the DAP series technical overview [2].

5.3.3   The Parallel Data Transform Library

In order to achieve efficiency it is important to map data in the appropriate way onto the DAP processor array. It is also often important to change the data mapping during processing. The execution of a DAP program can be thought of as a continuous cycle of activity [6], where each cycle consists of:

- Routing of data between the different PE's.

- Processing of local data on each PE.

The purpose of routing is to associate with each PE its designated operands before processing. Fortran Plus provides primitive functions which allow data to be shifted in each direction over the processor array. The user, however, often wants to perform more complex routing than this; for example, switching from horizontal to vertical storage. The Parallel Data Transform (PDT) library provides a method of describing a set of regular mappings; software is provided which allows the user to switch between these mappings efficiently. PDT's offer the following advantages over direct techniques [6]:

- Simplification of the description of data mapping and movement.

- Production of data routing code with faster execution time than with direct methods.

- Achieving data routing independent of the hardware routing network.

- Overcoming the constraints of fixed sized hardware.

Mapping Vectors

A mapping vector describes how the data is mapped onto the physical hardware. The mapping vector is used to transform the array index to generate a physical address:

$$\text{index} \longmapsto \text{physical address}$$

The mapping vector operates on individual bits of the index. These may be shuffled and inverted by the mapping vector, allowing a subset of regular mappings to be described. The subset is usually sufficient to provide most mappings required by the system designer. Efficient data routing is achieved by calculating a sequence of primitive routing stages to change between two mappings; this is performed in 'mapping vector space' rather than 'physical space', and code is then generated to perform the required series of data movements. A more detailed description of PDT's can be found in Flanders [52] and in Flanders and Parkinson [53].

Following this detailed description of DAP hardware and software, the next chapter describes how the DAP was used to implement the Ullmann subgraph isomorphism algorithm.

Chapter 6

The Ullmann Algorithm

6.1 Background

During previous studies [41] three techniques for detecting subgraph isomorphisms in small chemical structures have been investigated: set reduction [120], relaxation [126], and the Ullmann algorithm [124]. Downs *et al.* [41] showed that relaxation and the Ullmann algorithm utilise the same heuristic at different stages of a search, and, because the effectiveness of the heuristic is the most important factor in any such algorithm, they may be regarded as the same algorithm. It was also reported that the Ullmann algorithm was generally more efficient than set reduction for searching databases of small chemical structures [41]. The Ullmann algorithm was therefore chosen for a parallel implementation on the DAP.

6.2 Finding Isomorphisms

A subgraph isomorphism may be found by exhaustively comparing all subgraphs of a database structure with a query. Alternatively it might be possible to restrict the number of comparisons, if it is known that certain correspondences between query and database nodes are not permissible. For example, atom-types must match in chemical structures, and thus any isomorphism which maps an oxygen atom to a carbon atom will not be acceptable in the chemical context. Such heuristics can greatly reduce the number of comparisons that are required. Another important factor is the method of comparison between the query structure and database structure. One approach is to generate a database subgraph and then to test for an isomorphism with the query. A more efficient procedure is to build a mapping between query and database structures using some heuristic, stopping as soon as the mapping is recognised as being invalid; in this way mismatches may be identified before a complete mapping is generated. This procedure may be implemented by generating mappings by means of a depth-first or breadth-first search. In this case of substructure searching we wish to find a single subgraph isomorphism, and a depth-first search is thus appropriate. The heuristic is applied at each node in the search tree in order to eliminate branches leading to mismatches.

6.3   The Ullmann Heuristic

The heuristic assumes that an initial correspondence has been made between query and database nodes. Ullmann suggests a simple test for correspondence: a database node must have at least the same degree as a query node. Such a test can be used to limit mappings between a query and database node.

6.3.1   Terminology

Let the query and database graphs be represented by adjacency matrices A and B, respectively. Let the number of nodes in the query graph be P_α and the number of nodes in the database graph be P_β. Let M_{ij} be a matrix with $i=1\ldots P_\alpha$, $j=1\ldots P_\beta$ where

$$M_{ij} = \begin{cases} 1 & \text{if the } i\text{th node of } A \text{ can map to the } j\text{th node of } B \\ 0 & \text{otherwise} \end{cases}$$

The Ullmann heuristic may be stated as follows: given a query node a which has a neighbour b, and a database node x which can map to a, there must exist a neighbour of x, y, which maps to b. This can be expressed as:

if $V_{\alpha i}$ corresponds to $V_{\beta j}$ (i.e. $M_{ij} = 1$),

where $V_{\alpha i}$ is the ith node of A, $V_{\beta j}$ is the jth node of B

then

$$\forall x \ (1\ldots p_\alpha) \ ((a_{ix} = 1) \quad \Rightarrow \exists y \ (1\ldots p_\beta) \quad (m_{xy} \wedge b_{yj} = 1))$$

for all neighbours   there must exist   such that x and y

x of node i a node y in B correspond, and y is   (6.1)

a neighbour of node j

and if this is not true set m_{ij} to 0. The above heuristic is tested for every $m_{ij} = 1, i = 1\ldots P_\alpha$, $j = 1\ldots P_\beta$. Generally some of the m_{ij}'s will be negated, thus reducing the number of possible mappings between query and database nodes. If during this process any element of M is negated the refinement is repeated, since the negation may cause further changes to be made. The procedure is repeated until there is no change in M or until there is a zero row vector in M, which indicates a null mapping.

This algorithm can be described as a relaxation process. As discussed in Chapter 2, relaxation refers to a class of algorithms which solve many-variable problems by iteratively improving an initial solution using local information about a variable, until an acceptable solution is reached, or no further improvement can be made [126]. In this case a solution corresponds to a unique mapping for every query atom. The implementation of the refinement procedure will now be referred to as relaxation. It can be deduced that the relaxation procedure will be most effective when:

1. The database structure has low connectivity (B is sparse).

2. The average number of correspondences between query and database nodes is low (M is sparse).

This will improve the chances of

$$\not\exists y \ (1\ldots p_\beta) \ (m_{xy} \wedge b_{yj} = 1) = \text{TRUE}.$$

95

These conditions are satisfied for chemical graphs because:

1. The connectivity of a molecule is relatively low due to the limited valency of atoms.

2. The additional information accompanying chemical graphs (atom and bond-types) can be used to initially limit the number of correspondences between query and database nodes, which results in a sparse M matrix.

It therefore follows that the Ullmann algorithm is particularly suited to the processing of small chemical structures in 2-D, and this has been demonstrated previously in Sheffield [41].

If at any stage during the relaxation process the following condition holds:

$$\exists i(1\ldots p_\alpha) \text{ such that } m_{ij} = 0 \ \forall j \ (1\ldots p_\beta) \tag{6.2}$$

then a mismatch is identified since Equation 6.2 states that there exists a query node which does not map to any database node. This condition can be tested after every iteration of the relaxation process. The pseudo code for relaxation is:

$mismatch$:=FALSE;
REPEAT
  $change$:=FALSE;
  FOR i:=1 TO P_α
    FOR j:= 1 TO P_β
      IF (m_{ij}) THEN
      BEGIN
        $m_{ij} := \forall x \ (1\ldots p_\alpha) \ ((a_{ix} = 1) \Rightarrow \exists y \ (1\ldots p_\beta) \ (m_{xy} \wedge b_{yj} = 1));$
        $change := change \ \text{OR} \ (\text{NOT}(m_{ij}))$
      END;
  $mismatch$:=$\exists i(1\ldots p_\alpha)$ such that $m_{ij} = 0 \ \forall j(1\ldots p_\beta)$
UNTIL $(\text{NOT}(change) \ \text{OR} \ mismatch)$;

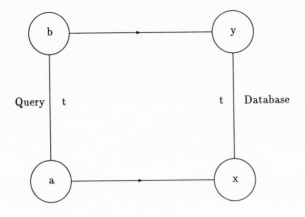

Figure 6.1: The Ullmann heuristic.

6.4 Implementation for Chemical Structures

The Ullmann algorithm processes general graphs and therefore does not take into account the bond information that is present in chemical graphs. The Ullmann heuristic utilises connectivity information only, and false drops may occur when it is used with chemical graphs unless a check is made to ensure the bonds connecting atoms match in the query and database structures. When using chemical graphs, bond information may be used in the heuristic by assuming that the matrices A and B contain integers representing bond types rather than Booleans for connectivity.

The Ullmann heuristic may then be restated: given a query node a connected to a neighbour b by an edge of type t, and a database node x; then, if a corresponds to x, there must exist a neighbour y connected to x by a edge of type t, such that y corresponds to b. This situation is shown in Figure 6.1. Equation 6.1 may be adapted to incorporate bond information and the resulting equation is:

$$\forall x \ (1 \ldots p_\alpha) \ ((a_{ix} = t) \ \Rightarrow \exists y \ (1 \ldots p_\beta) \quad (m_{xy} \wedge (b_{yj} = t))) \tag{6.3}$$

6.4.1 The Search Stage

The Ullmann algorithm proceeds using a depth-first search with the relaxation procedure applied at each tree node in order to eliminate branches leading to mismatches. In Ullmann's original paper the search process is implemented by iteratively manipulating the elements of the M matrix. The result is a complex algorithm which has been found to contain several errors in its original form (see Appendix A). The search algorithm described here is recursive, although the actual implementation was iterative: this was because iterative implementations are generally more efficient than the corresponding recursive versions and also because Fortran 77 lacks recursion.

```
CONST
    Pα=64; Pβ=64 {arbitrary limits};
TYPE
    maptype=ARRAY[1...Pα,1...Pβ] OF BOOLEAN;
VAR
    M:maptype;
    A:ARRAY[1...Pα,1...Pα] OF INTEGER;
    B:ARRAY[1...Pβ,1...Pβ] OF INTEGER;
    isomorphism:BOOLEAN;


PROCEDURE ULLMANN(d:1...Pα,M : maptype);
VAR
    M1:maptype;
    mismatch:BOOLEAN;
BEGIN
    M1:=M;
    REPEAT
        {choose new unique mapping for query atom d}
        {update M accordingly}
        relax(M,mismatch);
```

IF NOT(*mismatch*)

  IF $(d = P_\alpha)$ THEN

  BEGIN

    isomorphism:=TRUE;

    {output M as an isomorphism}

  END

  ELSE

    ULLMANN$(d + 1, M)$

  END;

  $M := M1$

UNTIL {all unique mappings for query atom d tried}

END;

BEGIN

  isomorphism:=FALSE;

  ULLMANN$(1, M)$

END.

6.4.2 Calculating the M Matrix

The matrix M is constructed so that:

$$
M_{ij} = \begin{cases}
1 & \text{if query atom } i \text{ is the same atom type as database atom} \\
& j \text{ and for every neighbour of } i \text{ there exists a} \\
& \text{neighbour of } j \text{ sharing the same atomic type and} \\
& \text{connected by the same bond-type} \\
0 & \text{otherwise}
\end{cases}
$$

This corresponds to testing whether the augmented atom i satisfies substructural equivalence with the augmented atom j. The condition usually results in a sparse M matrix unless the molecules are structurally similar and/or display symmetry.

The heuristic in Equation 6.3 involving bond-types is used in the relaxation procedure since 32-bit words are processed as efficiently as Booleans on a word-serial, bit-parallel computer such as the IBM 3083. Using Condition 6.3 will make the relaxation stage at least as effective as the connectivity case at no extra cost in processing time. A more effective relaxation stage will result in the elimination of branches leading to mismatches earlier in the search tree.

6.5 A Parallel Ullmann Algorithm

6.5.1 Approaches to Parallelism

Two types of parallelism can be explored in parallelising the Ullmann algorithm:

1. Algorithmic Parallelism: inherent parallelism within the algorithm is exploited.

2. Geometric Parallelism: perform many Ullmann searches together on different processors leaving the algorithm unchanged.

The first approach involves careful examination of the algorithm in order to discover which stages involve fine-grained parallelism. In this case we still match a single query against a single database structure and such a system could be easily integrated with existing software with the DAP acting as a search engine.

The second method is more complex; it involves matching a single query with several database structures in order to achieve high throughput. Geometric parallelism is generally effective on MIMD machines since work is simply distributed among available processors for independent execution, the only problem being that of load-balancing; this approach has been used in the substructure search system TOPFIT, to achieve near linear speedups for up to 30 processors [78]. On a SIMD machine, however, it is important that tasks are homogeneous in order that efficiency is maintained, since there is only a single instruction stream. Therefore the success of this approach would be crucially dependent on the level of synchronisation that could be achieved when matching a query with many database structures.

Because of the complexity of implementation and also the difficulty in predicting the performance of a geometric implementation, because of data dependence, it was decided to first examine an algorithmic methodology on the DAP.

6.5.2 An Algorithmic Technique

The algorithm described here will subsequently be referred to as Algorithm I. The algorithm may be divided into two stages:

1. A Tree Search;

2. Relaxation.

Tree Search

The method used here is depth-first since only a single subgraph isomorphism is required; this search method is inherently serial. Alternatively all branches from a node may be examined in a breadth-first manner. Such a brute force technique guarantees finding an isomorphism quickly (the problem has polynomial time complexity if infinitely many processors are available) since many mappings may be examined at once. This obvious algorithmic technique was suggested by Wipke and Rogers [137] for implementation in a MIMD environment. This approach is inefficient since the problem is now NP-complete in the number of branches to be tested; with many partial matches not leading to isomorphisms. For this reason it was decided that a depth-first search would be the preferred method of tree traversal.

Relaxation Stage

Condition 6.1 may be rewritten as follows:

$$\text{let } m_{ij} = m_{ij} \wedge \overline{r_i \wedge a_{ix}} \ \forall \ i,j \text{ such that } m_{ij} = 1$$

$$(6.4)$$

$$\text{where } r_i = \exists y \,(1 \ldots p_\beta) \quad (m_{xy} \wedge b_{yj} = 1)) \ (b_{yj} = b_{jy} \text{ since } b = b^T)$$

This is the reverse of Condition 6.1, and useful when considering parallelisation.

If A, B, M are DAP sized matrices, each element corresponding to a PE, r_i $(i=1\ldots P_\alpha)$ may be evaluated in parallel and the results placed in a vector r, thus

$$r_i \;=\; \exists y\,(1\ldots p_\beta) \quad (m_{xy} \wedge b_{yj} = 1)) \;\; \forall x = j \wedge x = 1\ldots p_\alpha$$

The matrices M and B must be aligned to allow this operation to be executed; this is done by working with M^T instead of M, to allow $m_{xy} \wedge b_{yj}$ to be executed as a single logical AND of two matrices. The vector r may be determined using the DAP function $ORCOLS$ which evaluates the OR of all elements in a matrix row for all rows in the matrix.

In order to execute a single relaxation, Condition 6.4 must be computed for $j = 1\ldots P_\beta$ given a fixed x. For the DAP implementation the reverse case is adopted: when executing Condition 6.4, x varies over $j = 1\ldots P_\alpha$, since we deal with the matrix A in parallel, and j is fixed by $x = j$. In order to vary j in the domain $j = 1\ldots P_\beta$, for a fixed x, M^T is shifted after each execution of Condition 6.4, P_β times in all. It also follows that the matrix A must be shifted similarly.

Let $MTRAN = TRAN(M)$ (the Fortran Plus function $TRAN$ evaluates the transpose of a matrix); then the Conditions 6.4 may be written in Fortran Plus as:

$R = MATR(ORCOLS(MTRAN.AND.B))$
(the function $MATR$ forms a matrix from a vector by replication)
$M = M.AND..NOT.((.NOT.R).AND.A)$

By De Morgan the second line may be simplified to give:

$M = M.AND.((R.OR..NOT.A)$

And the relaxation algorithm may be written as:

REPEAT

```
MTRAN = TRAN(M);
MCOPY = M;
FOR J:=1 TO Pβ DO
BEGIN
    R = MATR(ORCOLS(MTRAN.AND.B));
    M = M.AND.(R.OR..NOT.A);
    MTRAN = SHWC(MTRAN); {shift west cyclic}
    A = SHEC(A) {shift east cyclic}
END;
EXIT=.NOT.(ALL(ORCOLS(M)));
NOCHANGE = ALL(MCOPY.LEQ.M)
UNTIL (EXIT.OR.NOCHANGE)
```

The above code assumes we have a DAP with edge size P_β in order for cyclic shifts to route data in A and $MTRAN$ correctly (in a cyclic shift, data shifted from the PE's at the edge of the array, corresponding to the direction of the shift, is routed to those PE's on the opposite side of the array; in an acyclic shift the data is lost.). In practice $ES > P_\beta$, and one of the following strategies can be adopted:

1. Perform the loop ES times: this is optimally efficient when $P_\beta = ES$ but much less efficient when $P_\beta << ES$.

2. Replicate the data in A and $MTRAN$ and shift this data also in order to achieve the correct data alignment.

It was found that for all but the smallest values of P_β, the former approach was more efficient and this was the strategy used.

Tree Pruning

The effectiveness of a tree-pruning procedure can be quantified by the number of tree leaves which are eliminated. When optimising the performance of a tree-pruning

103

heuristic the shape of the tree is an important consideration. If the tree is very 'bushy' near the leaf nodes and sparse at the root, pruning at the top of the tree will eliminate many leaf nodes. However, if the reverse is true, pruning at the same levels will be less effective.

Thus it is important that the multiplicity of nodes high in the tree is minimised. This corresponds to matching query atoms which have few possible mappings onto database atoms. Before starting the search, the query atoms are sorted according to the number of tentative mappings each has with database atoms. The query structure is then renumbered so that the first atom has the least tentative maps and so on. The ensures that the search tree has the optimal shape for pruning. An experiment showed that on average such a renumbering yielded a 60% speedup. Consequently, both the serial and parallel codes were modified to utilise this approach.

6.5.3 Complexity

Serial Case

The serial relaxation may be considered as having complexity $O(P_\alpha^2 P_\beta)$ since Condition 6.1 must be checked for each m_{ij} and then for all a_{ix}.

Parallel Case

The parallel relaxation has complexity $O(ES)$ on an $O(ES^2)$ array processor since ES array operations are performed.

In practice only those $m_{ij} = 1$ are checked in the serial version and, since M is generally sparse, the serial algorithm is much more efficient than the complexity suggests. The parallel algorithm considers every element of M and therefore the speedup should increase as M becomes more densely populated. This corresponds to a decrease in the amount of information associated with the graphs before the search, as with, e.g., un-labelled, un-directed graphs. An experiment with two such fully connected graphs with $P_\alpha = P_\beta = ES = 64$ produced a speedup of 8000.

Bond/Connectivity Relaxation

In the serial case bond-types were considered during relaxation at no extra expense. On a bit-serial machine, however, the complexity of Condition 6.3 is $O(b)$ where b is the number of bits required to represent a bond-type. For example, if we require eight bond-types three bits are necessary to denote the bond and Condition 6.3 is tested for each bit. Thus the relaxation stage takes three times as long if bond-types are considered. This leaves two possibilities:

1. Relax using just connectivity information and check bond-types at the end of the search.

2. Consider bonds during relaxation, thus increasing the effectiveness of the heuristic.

Both approaches were tested with three bits used for the bond case.

6.5.4 Test Data

Five files of connection tables, each with an associated query, were used; the characteristics of which are shown in Table 6.2. The files were chosen so that the query substructure should be present in a high proportion of the structures comprising the database.

6.5.5 Initial Experiments

Initial experiments involved the use of Algorithm I to carry out searches for the parent substructure on five small sets of analogues from the medicinal chemistry literature, as described below. The fact that the query was contained in a large number of the file structures permitted verification that the algorithm executed correctly. Two sets of runs were carried out. In the first set, each relaxation procedure was executed with bond information; in the second, only connectivity information was used, with bonds checked at the leaf nodes. The results of these runs are shown in Table 6.3. $T(S)$ and $T(P)$ are the serial and parallel execution times respectively and $S(P)$ is the speedup or performance ratio; the figures in parenthesis are the number of searches per second.

An inspection of Table 6.3 shows that it is typically twice as fast to ignore bond information during the tree search stage. This can be explained by the limited valency of atoms and the low proportion of bond-types other than single or double. Most molecules contain many oxygen and carbon atoms which are generally bonded by just two types of bond, and therefore bond information is not very discriminating. The best speedups are obtained for large query and database structures as predicted by the complexity analysis. Consequently, subsequent experiments did not utilise bond information during the tree search stage.

6.5.6 Conclusions

It can be concluded that for the parallel case Ullmann is more efficient when connectivity information only is considered; this is a manifestation of the bit-serial nature of the DAP. In the serial program bonds are considered during the relaxation at no extra cost because complete words are processed in parallel.

The performance of the DAP is crucially dependent on the size of structures being processed. Performance is optimal when $P_\alpha = P_\beta = ES$ and poor when $P_\alpha \ll P_\beta \ll ES$. Thus when dealing with structures of less than 32 atoms a DAP 510 would be more efficient than a DAP 610. It is predicted that speedups will increase with P_α and P_β.

6.6 Data-Parallel Ullmann

6.6.1 Introduction

An alternative to algorithmic parallelism is that of data parallelism [30] in which many distinct items of data are processed concurrently on separate processors. This is sometimes referred as as outer-loop parallelism [110] since concurrency occurs

106

at the outer level. In this section a data parallel implementation of the Ullmann algorithm is described; this will subsequently be referred to as Algorithm II.

6.6.2 Data Mapping

The objective of the data-parallel approach is to match many structures with a single query structure simultaneously, to achieve a high throughput. An obvious way to achieve this is to assign structures to PE's, with the query structure stored centrally. This implies storing all the data associated with individual structures in the local memory of a PE, which necessitates the use of a vertical mapping for structure data. In the DAP used for these experiments each PE has a local memory of 64 Kbits, which is a relatively small amount. Therefore care had to be taken to minimise memory usage.

6.6.3 Algorithm Design

There are two problems associated with the design of a data parallel algorithm on a SIMD processor:

1. Correct local execution;

2. Global control to ensure 1.

In a data parallel implementation the original algorithm remains intact and both the algorithm code and data are replicated on processors in order to effect parallel execution. Therefore the first task is to produce an efficient version of the algorithm for execution on a single processor. Once this has been done a control mechanism must be designed that ensures that all processors execute the algorithm correctly and efficiently. On a MIMD machine the control problem is simplified since each processor has a high degree of local control in hardware. On a SIMD machine, where the control is centralised, the goal is to achieve optimal synchronisation.

107

Local Execution

One must consider the implementation of the algorithm on a single DAP PE, in isolation, executing instructions from the MCU. The task is to establish a depth-first search algorithm, with a tree-pruning heuristic, on a bit-serial processor which may only perform single-bit logic and arithmetic. Furthermore, memory is limited to 64 Kbits and all memory access must be performed one bit at a time. Although a high-level language is used, this effectively acts as an interface to low-level instructions. Any use of 'high-level' data types will result in a loss of flexibility and performance.

The first step was to achieve an implementation of Ullmann which executed on single PE's correctly without regard to global control. For this purpose, it is assumed that every PE has exclusive access to the MCU. But since there is only a single MCU, this implies that each PE must execute the same instructions, and consequently each PE must have the same local data to ensure correct execution. A control mechanism was designed at a later stage which permitted the variation of control based on local data, despite the absence of local control hardware.

To achieve the required parallelism and low-level flexibility, all data was stored as matrices of logical variables. The original algorithm was then adapted with scalars being replaced by matrices and non-logical data replaced by linear arrays of logicals. This implementation was then tested with a variety of query and database structures to ensure that the basic search mechanism was correct. This was a time-consuming process, since visualisation of large numbers of logicals is difficult. To aid data visualisation, a series of PDT's were written which remapped the data into more convenient forms. After the code had been fully tested the design of the global control mechanisms was started.

Global Control

The Ullmann algorithm may be broken down into a sequence of discrete stages. Figure 6.2 shows the sequence of execution of these stages. The path generated by each structure will generally be unique because of the diverse topologies and node

108

labellings of the structures in the database. Since there is not enough hardware control in a PE to implement the correct execution sequence, this must be handled by the MCU. The MCU can only issue a single instruction stream, and each individual PE must therefore select only a subset of these instructions based on local conditions. In this way each PE can execute the correct set of instructions irrespective of the global sequence.

This strategy may be implemented by using logical masking: each step has a logical mask associated with it; if a PE elects to execute an instruction its mask will be set to TRUE, otherwise it will be set to FALSE. The logical mask is used to set the activity register within the PE. The masks are updated according to the sequence of steps required to ensure correct execution of the algorithm. This method provides an efficient way of providing the correct local control, irrespective of the global control path supplied by the MCU.

Once such mechanisms were implemented it was possible to test the algorithm with 4,096 structures. The code was tested by comparing the results of the 4,096 different parallel searches with those produced by the serial algorithm.

Optimising Global Control

During the initial runs testing was achieved by the MCU broadcasting the instructions for each step sequentially, i.e., step1, step2,....., step7, step1,..., repeatedly until all PE's had completed the algorithm (where step n denotes the nth step in Ullmann's original algorithm, see Appendix A); we shall refer to this subsequently as the arbitrary step selection method. Although such an arbitrary global sequence is guaranteed to work eventually, it will not necessarily do so in the shortest time. To minimise the execution time it is essential to maximise the PE utilisation. A bad global sequence will leave many PE's idle because of local inactivity, i.e., bad synchronisation. The ideal is to choose a global sequence which synchronises the local paths of PE's, in order that many PE's will elect to execute the same instructions together. The best path will vary for different queries and datasets, and is obviously

109

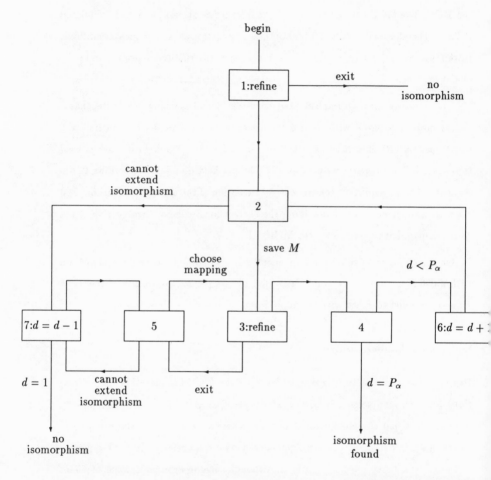

Figure 6.2: Flow of control in the Ullmann algorithm. The numbers in each box correspond to the main steps of the algorithm as detailed in Ullmann's original paper (which is described in Appendix A); d is the current level in the search tree.

very data dependent. For this reason the execution sequence is chosen dynamically, as the algorithm executes.

Majority Voting

The majority voting method assures optimal synchronisation for the next step in the global path. When deciding which step should be executed, a poll is taken of all the possible choices. The most popular step is then chosen for execution. This scheme is simple and has an efficient implementation. The TRUE positions in each logical mask are summed to give a scalar value: this value represents the number of PE's wishing to execute that step. The summing operation may be executed very efficiently by a parallel algorithm with logarithmic complexity.

The majority voting control method was about an order of magnitude faster than the arbitrary step selection method. Although the method achieves optimal synchronisation for the next step, it does not ensure optimal synchronisation for the complete algorithm. The problem of determining the best sequence of steps before execution is analogous to the scheduling problem [60], in which the order of execution of a number of tasks, requiring varying resources, is to be calculated. The scheduling problem is generally NP-complete and thus intractable; for this reason it was felt that the performance of the majority voting method, although not optimal, was acceptable.

6.6.4 Relaxation Procedure

For Algorithm I it was shown that at least 90% of the total execution time was spent performing relaxation operations. Therefore it was important to design an efficient code for this part of the algorithm.

In the data parallel implementation of the relaxation procedure, the heuristic is checked for every element of M, whether set or not. This is because it is likely that when dealing with 4,096 structures at least a single bit will be set in each plane i, j. Thus the data parallel implementation would be more efficient for the processing of dense M matrices, than sparse M matrices, as is the case with the algorithmic approach.

Despite this unavoidable inefficiency, the data parallel algorithm is efficient in the checking of the heuristic for neighbours of the query atom i. Rather than check the heuristic for every possible neighbour x of atom i, the algorithm checks only those bonded neighbours. This is because only a single query, for which the topology is known explicitly, is being compared with the database.

For efficiency reasons it is convenient to store the query as a sequence of pointers rather than an adjacency matrix. The data structure used is an array of integer vectors; each vector contains pointers to the neighbours of an atom. Although this mapping is wasteful in terms of storage, it facilitates efficient memory access, since the pointers index M directly. Thus the updating of M can be performed extremely efficiently.

6.6.5 Tree Compression

Storage of the search tree dominates the storage requirements of the Ullmann algorithm. For a depth-first search, d mappings must be stored (where d is the tree depth). Each mapping is represented by an M matrix of size $P_\alpha * P_\beta$ and $d = P_\alpha$ (the number of atoms that must be assigned). Therefore the storage required is $O(P_\alpha^2 P_\beta)$ bits. Table 6.1 shows the storage requirements as P_α increases, with $P_\beta = 64$. Given the DAP available to us, the only feasible choices are $P_\alpha = 16$ or 24. In order that sufficient memory was left for unforeseen requirements a maximum query size of 16 atoms was chosen. This seemed reasonable since it is unusual for a chemist to use very large query structures. In an operational system more array store would be available and no such restrictions would apply.

The data for each structure consisted of an M matrix of dimensions 16*64 bits and a bond adjacency matrix of dimensions 64*64*3 (three bits for each bond, see Section 6.5.3). Each was mapped vertically in the local memory of a PE, with the first dimension changing most rapidly.

P_α	Search Tree Size (Kbits)	
	Without Compression	With Compression
16	16	8
24	36	12
32	64	16
48	144	24
64	256	32

Table 6.1: Relation between maximum query size and memory requirements.

As the development of the algorithm proceeded it soon became apparent that more array store would be required. The largest data structure, in terms of memory consumption, was the search tree, which occupied a quarter of the available array store in each PE.

As described earlier in this section, the search tree is represented by storing P_α copies of the M matrix, one for each level of the tree, the resultant memory requirement being $O(P_\alpha{}^2 P_\beta)$. This may be reduced if one recognises the following: if a tentative mapping is not possible at the top of the tree, it will not exist elsewhere in the tree. In terms of the M matrix this means that as the search proceeds, down the tree, a TRUE element may change to a FALSE (when a mismatch is found), but not vice versa. Thus it is only necessary to store the tree level at which a TRUE element is negated. An 8-bit integer is used for this purpose and the storage requirement, when tree compression is used, is reduced to $O(8P_\alpha P_\beta)$.

Since the storage complexity for this method replaces a P_α term with a constant, the demand for storage does not increase with the query size as rapidly as with the former method. Furthermore, an 8-bit integer allows up to 256 search tree levels to be stored: equivalent to a query structure of 256 atoms. Table 6.1 shows the relationship between query size and memory requirement, for a fixed $P_\beta=64$

113

atoms, when tree compression is used. For the experiments, the maximum query size remained fixed at 16 atoms. Since previously this required 16 Kbits, 8 Kbits was made free in each PE for other purposes.

6.6.6 Algorithm Inefficiencies

Data Precision

Most code sections in the algorithm have a complexity term P_β, in other words the time taken to execute many parts of the algorithm is dependent on the size of the molecules to be processed; indeed the relaxation procedure has complexity $O(P_\beta^2 P_\alpha)$, making the execution time even more susceptible to the structure size of the molecule. Since many structures are processed simultaneously the molecule with the largest number of atoms will determine the complexity for a particular step in the algorithm. Thus if a single molecule has 64 atoms and the rest less than 20 atoms, the execution time will be the same as for a dataset consisting entirely of 64 atom molecules. This inefficiency is unavoidable and is caused by the lack of local control within DAP PE's. However, the extent to which this inefficiency affects execution time may be controlled in two ways:

1. Choose datasets which have a low distribution of structure sizes. This could be achieved by sorting the database in terms of molecule sizes. Unless the database was extremely large, however, each block of 4,096 molecules would still contain some disparity of molecule sizes.

2. Before a code section is executed, determine the size of the largest structure wishing to partake in that step; then execute the code section to that precision, for example:

$P_{\beta max} = MAXV(PBETA, STEPMASK)$
FOR $index := 1$ TO $P_{\beta max}$ DO
$\quad$ {code section}

114

where,

$PBETA$ is a matrix of molecule sizes

$STEPMASK$ is the logical mask for the step to be executed

The effects of the first optimisation were tested by using datasets which had a variety of ranges of structure size and the second optimisation was incorporated into all parts of the source code for Algorithm II.

Finishing Sequence

After a PE finishes its search it remains idle until all the other PE's have completed their searches also. This effect is minimal when all the structures finish together. Often a group of molecules, however, or even a single molecule, will take considerably longer than the rest to process. Under these circumstances the PE utilisation is low, since many PE's remain dormant for long periods. This effect was measured and the results are discussed in Section 6.7.3. If large numbers of molecules are available for processing the effect can be reduced; an algorithm for achieving this is described in Section 6.9.

Tree Pruning

Unfortunately, in the data parallel case the search trees could not be optimised for pruning as with the serial and algorithmic parallel programs (see Section 6.5.2). This is because the renumbering of the query is based upon a comparison of the query and database structures. When dealing with many database structures many conflicting renumberings result. Since the query is stored centrally, only a single renumbering is possible.

Bond Checking

The initial experiments (see Section 6.5.5) demonstrate that it is quicker to check bonds at search tree leaves rather than at every node, for the algorithmic parallel

115

program. This approach is not possible in the data parallel case since, at any stage in the search, different PE's will be at different places in the tree, with a high probability that at least a single PE will be at a tree leaf and thus require a bond check. One approach is to wait until all PE's are synchronised at a leaf and then apply the expensive bond check. This results in poor synchronisation, however, because many PE's will be forced to wait for the others to 'catch up' and reach a leaf. For this reason it was decided to apply bond checking throughout the tree. Although this results in a three-fold inefficiency (since the relaxation procedure must be invoked for each bit representing the bond), it will increase the effectiveness of the tree pruning heuristic.

6.7 Main Experiments (Algorithms I & II)

6.7.1 Database Structures

The main experiments were performed with seven sets of 4,096 structures from the Fine Chemicals Directory (FCD) [127]. The FCD is an index of commercially available compounds compiled from suppliers' catalogues. The version of FCD available at Sheffield University contains about 60,000 molecules in the WLN format. The DARING program [43] was used to convert the WLN's to redundant connection tables for the experiments.

The seven data sets consisted of three random samples. The other four sets were chosen according to the range of structure sizes within the set; for this purpose a frequency distribution of the structure sizes within FCD was constructed. Three of the sets consisted of samples with a very low variation of structure size; it was thought such a sample would benefit Algorithm II, as discussed in Section 6.6.6. The other set was the set of 4,096 largest structures from the FCD with $P_\beta \leq 64$: this set was chosen in order to test whether Algorithm I performed more efficiently than the serial program with large structures, as predicted by the complexity analysis. Details of each of the seven data sets are given in Table 6.4.

116

6.7.2 Query Structures

64 query structures were chosen from the journal literature on substructure searching algorithms and substructure searching systems. The structure diagrams for these molecules are shown in Appendix B. Of these, seven contained more than 16 atoms, and thus could not be used with Algorithm II. Results for these queries are presented for comparison with the serial algorithm. The mean number of atoms, with the standard deviation in brackets, for the set of 57 queries and the remaining seven were 10.07 (3.37) and 19.28 (2.14) respectively.

6.7.3 Results

The run-times for the seven data sets are given in Table 6.5. The reader will notice that the relative performance of Algorithm I increases with the mean value of P_β, as predicted by the complexity analysis: the second worst speedup is obtained for dataset 7 which has the lowest mean structure size; similarly the best speedup is achieved for dataset 6 which contains the largest molecules. Algorithm II equals or outperforms Algorithm I for only two of the seven datasets, these being datasets 4 and 5. These are the samples with the lowest distribution of structure sizes; indeed, the speedup value shows an inverse relation to the standard deviation in structure sizes for all of the datasets.

Although the speedups obtained may seem low it is important to remember that the serial implementation of the algorithm has an extremely efficient implementation. This is demonstrated by the processing rates, which in many cases exceed 1,000 searches per second.

It should be noted that the results in Table 6.5 are cumulated over a large number of searches: 233,472 in all (57 queries, each matched against a database of 4,096 structures). Although Algorithm II is very efficient for some searches, the problem is that the overall run-time is determined by the time for the slowest match. Thus, even though a large number of the 4,096 structures may have completed, a small number of long-running matches, or even a single slow match, can result in

117

the algorithm giving an extended run-time for a particular query. This effect can be measured by counting the number of molecules that finish after each call of the relaxation procedure. Thus, for one of the queries, the first seven calls of the relaxation procedure resulted in the completion of the processing for 0, 0, 208, 3178, 578, 100 and 9 molecules; calls 8-41 produced no further completions and then calls 42-44 resulted in the completion of 6, 2 and the final 5 molecules. Again, for another query, calls 1-4 of the relaxation procedure resulted in the completion of 0, 0, 4040 and 35 molecules; calls 24, 25 and 30 produced 12, 1 and 7 completions, respectively, with the single remaining molecule completing only after call-55. A better example of the utility of Algorithm II is provided by a query where all of the molecules were processed by a total of just seven calls to the refinement procedure, these resulting in the successive completion of 0, 0, 3411, 683, 0, 1 and 1 molecules.

A measure of the overall efficiency of Algorithm II is given by the percentage utilisation of the PE's, which may be ascertained by noting how many PE's are active during each call to the refinement procedure. The mean utilisation for the seven datasets, with the standard deviations, minimum, maximum values and the total number of invocations of the refinement code are shown in Table 6.8. The reader will note the great differences between the minimum and the maximum values (and similarly the very large standard deviations): these serve to emphasise the differences that have been exemplified in the previous paragraph. Since the mean efficiencies are comparable, the differences in the performance ratio that are observed in Table 6.5 must be accredited to other factors. It is believed that the main factor is the inefficiency caused by a disparate distribution of structure sizes within a file, as discussed earlier.

It is important to note that the precise performance ratios that have been obtained with the sets of FCD structures may be significantly lower than those that would be obtained in an operational substructure searching system, where an initial screening search would be used to eliminate many of the database structures from the time-consuming atom-by-atom search. The sets of 4096 molecules used here contained very many structures that had little or no resemblance to the query substructures that were searched. These structures were thus eliminated at a much earlier stage of the processing than would be the case with conventional screened output. In this respect, the performance ratios obtained with the small datasets may give a better indication of the efficiency of the parallel algorithms, since most of these structures were analogues having a large degree of commonality with each other and with the query substructure (as would normally be the case with the output from the screening stage of a substructure search). It is thus likely that the efficiency of the parallel processor, relative to the sequential processor, increases with the amount of computational work that is required. Evidence to support this suggestion comes from searches of the seven datasets with queries for which $P_\alpha > 16$, and which could thus not be searched using Algorithm II. The results for these searches are shown in Table 6.7. The reader will note that the speedups are significantly better than those obtained for the smaller queries. It will thus be clear that the relative performances of the various algorithms are crucially dependent on the particular data that is to be processed.

The low processor utilisation in Algorithm II suggests the need for a mechanism to decrease the number of inactive PE's: this point is addressed in the two algorithms that are discussed in the remainder of this section.

6.8   Hybrid Algorithm

The deleterious effect of long-running molecules on Algorithm II suggests two further algorithms that could be adopted. The first of these combined algorithms, referred to subsequently as Algorithm III, involves two stages as follows:

- Continue the execution of Algorithm II until some user-defined number, n, of the matches have been completed.

- The remaining 4096-n structures (or 1024-n structures for a DAP 510) are then processed in sequence using Algorithm I.

Tests with this integrated strategy suggest that its overall efficiency is dependent on the value of n. In the experiments n was increased from 2,000 to 4,000 in increments of 50. The optimal value of n, referred to subsequently as the cut-off, was the value which resulted in the shortest run-time. Where the shortest run-time was achieved for several values of n, the largest value of n is reported as the cut-off.

The results for the hybrid algorithm are shown in Table 6.6. The performance of the hybrid algorithm is in all cases superior to the serial and parallel implementations I and II. This suggests that the combined approach achieves a higher degree of PE utilisation than Algorithms I and II. The cut-off values for each dataset are comparable and this indicates that the number of difficult, or long-running molecules in a sample is in general about the same, irrespective of the other characteristics of the dataset.

It should be noted that a lack of array store prevented the preserving of search information gained in the first stage of the search, for those $4096 - n$ structures which had not finished searching, for use by Algorithm II. In an operational system, the search trees, together with information about the current state of the search, could be passed to the second stage in order to avoid redundant processing. In the experiments, Algorithm II started all searches from the root node of the search-tree. As a result of this the search times presented for Algorithm III are slower than those that would be obtained in an optimised system.

6.9   Algorithm IV: The Drip-Feed Algorithm

The lack of efficiency in Algorithm II, discussed in Section 6.6.6, can be avoided if large numbers of molecules are available for processing. The approach uses the processor-farm technique, in which work is dynamically distributed to a 'farm' of processors:

1. Continue the execution of Algorithm II until n of the matches have been completed.

120

2. Suspend processing and load n new structures into those PE's where processing has been completed.

3. Repeat Steps 1 and 2 until all of the structures that are to be searched have been completed.

The complexity here lies in the need to re-fill the PE's rapidly; this is achieved by using a PDT to route 'fresh' molecules to those PE's which are idle. The parallelism is maximised with a small value of n but this is at the cost of the frequent loading of new structures; alternatively, if a large value of n is used, the efficiency of the algorithm will approach that of Algorithm II.

At present the searching of such large databases is an unusual requirement, and not worth going to the extremes of effort and resource necessary to acquire and manipulate an experimental database of this size. However, it is believed that on very large databases algorithm IV would offer the highest level of performance of the four algorithms presented.

6.10   Conclusions

In this chapter several novel parallel implementations of the Ullmann subgraph isomorphism algorithm algorithm have been described. The first, Algorithm I, used algorithmic parallelism to perform a fast match between a query and database molecule. Algorithm II utilised data-parallelism, and although much slower than Algorithm I in operation, performed 4,096 matches simultaneously.

It was found that Algorithm II was potentially the fastest algorithm. In the experiments, however, a small percentage of long-running matches reduced the overall

121

run-time. A hybrid system, comprising the algorithmic and data-parallel programs, overcomes this difficulty: the combined approach is faster than either algorithm I or II.

Until now we have only been concerned with the atom-by-atom stage of a substructure search. This was because the original intention was for the DAP to be used as a substructure search engine, with all other operations performed on the host computer. It was found, however, during the course of the experiments reported in this chapter, that this approach was not feasible when processing large databases, because of inefficiencies in the host-to-DAP data-transfer system. For this reason a substructure system running entirely on the DAP, with little host intervention, was designed; this work is reported in the following chapter.

Dataset	P_α	P_β			Comments
		Mean	SD	Maximum	
A	10	11.9	1.2	16	28 benzohydroxamic acids [125]
B	9	25.3	8.6	36	105 benzamidines [66]
C	13	17.2	1.9	20	24 hydrazides [112]
D	12	26.4	2.9	37	79 penicillins [8]
E	20	29.6	4.9	37	46 quinazolines [32]

Table 6.2: Characteristics of the five small datasets from the medicinal chemistry literature that were used in the initial experiments.

Dataset	$T(S)$	$T(P)$		$S(P)$	
		Connectivity Relaxation	Bond Relaxation		
A	0.56 (50)	0.10 (280)	0.21 (133)	5.37	2.67
B	7.93 (13)	0.90 (117)	1.60 (65)	8.85	4.96
C	1.87 (13)	0.25 (96)	0.45 (53)	7.47	4.16
D	2.28 (35)	0.49 (161)	1.19 (66)	4.69	1.92
E	3.56 (13)	0.44 (105)	0.92 (50)	8.15	3.87

Table 6.3: Execution times (in CPU seconds) on an IBM 3083 ($T(S)$) and on a DAP 610 ($T(P)$), and speedups ($S(P)$) for the use of Algorithm I on five small datasets from the medicinal chemistry literature. The first of the two figures for $T(P)$ and $S(P)$ correspond to those runs of the refinement procedure that used just the connectivity information during the refinement, with the check on bond-types being executed once all of the query atoms had been mapped, and the second of the two figures to those runs that used the actual bond-types during the refinement. The number of searches per second is given in parenthesis.

123

Dataset	P_β			Comments
	Mean	SD	Maximum	
1	25.92	6.58	64	Random set
2	16.23	7.33	64	Random set
3	22.00	1.68	25	Centre of the distribution
4	16.47	0.56	17	$P_\beta= 16,17$
5	21.48	0.60	22	$P_\beta= 21,22$
6	37.46	7.71	64	$P_\beta> 30$
7	16.03	6.94	61	Random set

Table 6.4: The mean, standard deviation, and maximum structure size for each of the seven data sets used for the main experiments.

Dataset	$T(S)$	$T(P)$		$S(P)$	
		I	II	I	II
1	304.93 (766)	186.45 (1252)	573.24 (407)	1.64	0.53
2	137.47 (1698)	138.94 (1680)	451.86 (517)	0.99	0.30
3	215.26 (1085)	178.88 (1305)	212.79 (1097)	1.20	1.01
4	160.46 (1455)	173.64 (1345)	121.17 (1927)	0.92	1.32
5	206.77 (1129)	179.25 (1302)	179.73 (1299)	1.15	1.15
6	477.94 (488)	172.75 (1352)	822.98 (284)	2.77	0.58
7	121.83 (1916)	137.05 (1704)	354.59 (658)	0.98	0.34

Table 6.5: Execution times (in CPU seconds), with the number of searches per second in parenthesis, and speedups ($S(P)$) for the serial implementation of Ullmann's algorithm and the two parallel implementations I and II, when matching the set of 57 queries with $P_\alpha \le 16$.

Dataset	$T(S)$	$T(P)$ III	$S(P)$	Cut-off
1	304.93 (766)	138.93 (1681)	2.19	3600
2	137.47 (1698)	122.57 (1905)	1.12	3800
3	215.26 (1085)	46.41 (5031)	4.64	3700
4	160.46 (1455)	30.83 (7573)	5.20	3850
5	206.77 (1129)	43.72 (5340)	4.73	3800
6	477.94 (488)	150.04 (1556)	3.19	3300
7	121.83 (1916)	110.41 (2115)	1.10	3900

Table 6.6: Execution times (in CPU seconds), with the number of searches per second in parenthesis, speedups ($S(P)$), and cut-off value for the serial implementation of Ullmann's algorithm and the hybrid algorithm III, when matching the set of 57 queries with $P_\alpha \le 16$.

Dataset	$T(S)$	$T(P)$	$S(P)$
1	233.11 (123)	57.69 (497)	4.04
2	42.76 (671)	12.06 (2377)	3.55
3	89.29 (321)	30.39 (943)	2.94
4	2.98 (9621)	2.64 (10861)	1.11
5	75.49 (380)	27.47 (1172)	3.09
6	302.53 (95)	62.05 (462)	5.52
7	18.74 (1530)	6.99 (4102)	2.68

Table 6.7: Execution times (in CPU seconds), with the number of searches per second in parenthesis, and speedups ($S(P)$) for the serial implementation of Ullmann's algorithm and the parallel implementation, Algorithm I, when matching the set of seven large queries with $P_\alpha > 16$.

Dataset	PE Utilisation				Refinement Invocations
	Mean	Standard Deviation	Minimum	Maximum	
1	9.67	40.85	1.73	94.53	2706
2	10.65	33.06	3.37	93.16	2297
3	11.03	33.54	1.70	95.32	2376
4	11.54	34.27	4.39	94.82	2185
5	10.83	34.46	1.73	95.09	2409
6	10.66	39.15	2.20	96.04	2520
7	12.61	34.36	4.18	94.46	1941

Table 6.8: Percentage mean utilisation of PE's, with standard deviation, minimum, maximum values and total numbers of invocations of the refinement procedure for each of the seven datasets used in the main experiments.

Chapter 7

A DAP-Based
Substructure-Search System

7.1 Introduction

The subgraph matching algorithm has been described in detail in Chapter 6. In this chapter the design of a complete substructure-search system, running on the DAP, is reported.

7.2 Background

A substructure search system involves several stages of processing, if we assume a file of structures that have passed the screening stage. These are:

1. Read in structure data from backing store.

2. Transform structure representation to one suitable for the search algorithm.

3. Create search space.

4. Perform search.

5. Output results.

All but stage 4 may be considered tractable in terms of complexity. That is, it is possible to find an efficient serial implementation. It is the actual matching procedure that is the computationally demanding phase, and it has been found that this is the computationally dominant part of substructure searching systems. The object of this study is to find an efficient parallel implementation of the matching procedure, it being assumed that fast procedures for the other stages have already been found. In this thesis, it is the subgraph isomorphism procedure that is used for the comparison between serial and parallel algorithms.

7.3 System Design

First, a completely serial substructure search system was developed for a fast serial computer, in this case Sheffield University's IBM 3083 BX mainframe. This permitted familiarisation with the algorithm before attempting to code the more complex parallel version. When designing the serial code, most attention was directed towards designing an efficient matching stage. This is very important if a fair comparison between parallel and serial algorithms is to be made. Time did not permit the design of optimised code for the other stages; the objective being to produce a system which generated data for the search stage quickly enough for testing purposes. Such an approach permitted some of the serial code to be re-used in the DAP system. Specifically, the matching stage could be replaced with the appropriate parallel code running on the DAP, leaving the other stages intact. Figure 7.1 shows the configuration of the substructure search system.

7.4 Structure Input

The performance of the serial system was slow in comparison with commercial systems, but adequate for research purposes. This was because only the matching algorithm was optimised, and not the other components of the system. The same was true of the parallel system when small files of structures were being searched;

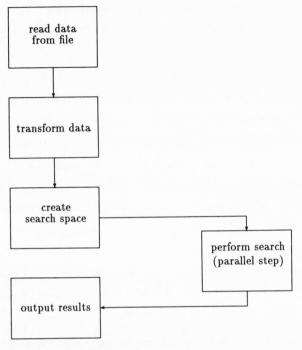

Figure 7.1: Configuration of a substructure search system.

an inefficiency became apparent, however, when larger files were processed.

The input data for the Ullmann algorithm is most effectively stored as arrays of logical variables. Typically, each structure occupies several thousand bits of storage space. This data must be transferred from the host computer to the DAP and then transformed into a representation recognised by the DAP (see Section 5.3.2). The problem is that logical variables are stored as whole words on the VAX host, i.e., a single bit occupies 32 bits of storage space. This results in a 32-fold inefficiency when transferring logical data from the host to the DAP, and makes the system prohibitively slow for the processing of all but the smallest of files.

In the first case, the DAP system was used to test the algorithmic version of the Ullmann algorithm with files of up to 100 structures. The system completed the searches in minutes, which was fast enough for program development. For the data-parallel algorithm, which performs 4,096 searches simultaneously, development was undertaken with only 64 structures because of the difficulty in transferring large numbers of structures in a reasonable time.

It was realised that a system with greater throughput would be required after the initial development stage in order to perform effective testing and optimisation for both parallel algorithms. It was obvious that a radically different system architecture would be needed to overcome the I/O difficulties encountered. One approach was to read data directly into the DAP and perform all of the necessary pre-processing on the DAP. This was attractive since little re-coding would be required. It was found, however, that the serial code performed very badly on the DAP. This was because only a single record could be read from the disk in a single time-slice (the DAP being a multi-user system). The combination of a relatively long time-slice, and a short record length, meant that the absolute time required to read a single structure from backing store was lengthy when there was more than one DAP process being executed.

Rather than waste more time attempting to find partial solutions to the I/O problem, it was decided to re-design the DAP system from first principles, attempting to optimise the system for parallel execution on the DAP.

7.5 Design of a Parallel Search System

7.5.1 Data Storage

The first consideration is the manner in which structures are represented. The most common method for use in substructure search systems, and the one adopted for this work, is the redundant connection table (see Section 2.3.3). The scheme is expensive in terms of storage, but its use has become more widespread as the price of backing store has reduced. Consider the connection table shown in Figure 7.2; each line of the connection table is stored as a fixed length record (Fortran cannot read files of

```
****************************************************************

STRUCTURE NUMBER 1039

    1    O    0    1    2(1)    5(1)
    2    C    0    0    1(1)    3(2)
    3    C    0    0    2(2)    4(1)
    4    C    0    0    3(1)    5(2)
    5    C    0    0    4(2)    1(1)


****************************************************************
```

Figure 7.2: Format of a connection table generated by DARING.

variable length records) of 62 characters. Before and after each structure are several blank lines, with an additional line of stars delimiting structures in the file; within each line, multiple spaces delimit fields. The connection table is very wasteful of storage: in the example, formatting characters account for about three quarters of the storage and, furthermore, each connection is stored twice. The inefficient utilisation of storage is justified by the ease of data visualisation which the scheme allows. That is, it is very easy to generate a structure diagram from the connection table.

It was decided to design an intermediate binary representation of the structure file which would minimise data storage and allow efficient processing by the search algorithms. The structure of the binary representation closely follows that of the non-redundant connection table. Storage is minimised by adopting variable length records for the storage of each atom and its neighbour connections. Further reductions are achieved by using the minimum precision, in bits, to store each field, as shown in Table 7.1. The total storage, in bits, required by a structure, is given in Equation 7.1.

131

Field Name	Storage (bits)	Explanation
Atom count	7	Number of atoms in the structure
Atomic number	7	Atomic number of the atom
Neighbour count	3	Number of bonded atoms
Neighbour number	7	Label of neighbour
Bond type	3	Type of bond connecting atom to its neighbour

Table 7.1: Data fields used in the binary representation of a connection table

$$\text{Storage (bits)} = 7 + \sum_{i=1}^{Atom\ count} \left(10 + \sum_{j=1}^{Neighbour\ count(i)} 10 \right) \tag{7.1}$$

Example Binary Coding of a Structure

Consider the following connection tables:

Redundant Form

1	O	2(1)	5(1)
2	C	1(1)	3(2)
3	C	2(2)	4(1)
4	C	3(1)	5(2)
5	C	4(2)	1(1)

Non-Redundant Form

1	O		
2	C	1(1)	
3	C	2(2)	
4	C	3(1)	
5	C	4(2)	1(1)

Bit-String Representation:

```
    5
0000101
    O           0
0001000        000
    C           1          1          1
0000110        001      0000001      001
    C           1          2          2
```

132

0000110	001	0000010	010		
C	1	3	1		
0000110	001	0000011	001		
C	2	4	2	1	1
0000110	010	0000100	010	0000001	001

The structure can thus be represented by the bit-string:

0000101000100000000001100010000001001000011000100000100100000110001000
0011001000011001000001000100000001001

This occupies 107 bits, or 14 bytes, compared with 496 bytes for the original connection table, giving a compression ratio of 1:35.

7.6 Generating Binary Structure Files

Having decided on a binary structure representation, it is necessary to develop software to compress existing structure files into binary files. This process involves extensive bit-level processing and therefore requires an environment which supports such low level processing. The DAP is ideal for this task since processing is performed at the bit-level and Fortran Plus provides many facilities for bit-level manipulation.

It is not important to produce an efficient program for file compression since the compression stage is executed only once and not in real-time. The application, however, is naturally data-parallel, i.e., the same operation is to be applied many times in succession, with each structure being independent of the rest of the file. For this reason a parallel data compression algorithm was chosen.

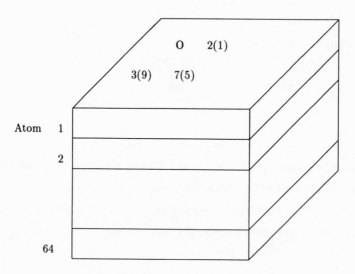

Figure 7.3: Horizontal storage of structures.

7.7 Compression Algorithm

7.7.1 Data Input

Given an ASCII representation of a connection table line, the task is to extract the relevant information and transform this into an equivalent compressed binary code. Each connection table line contains a variable number of characters, dependent on the number of atom neighbours. Each line, however, has the same format and it is therefore appropriate to assign lines to PE's; this implies a vertical mapping of the structure data on to the array store.

The first stage of the algorithm involves reading data from the structure file into array store. The system routines provided for this process assume that a horizontal mapping is required. Records are read line-by-line, with each record occupying a single storage plane. Assuming a maximum structure size of 64 atoms, 64 structures can be compressed simultaneously. After a 64-structure block has been read into array store, a PDT is used to map atoms on a PE basis. Each structure now

134

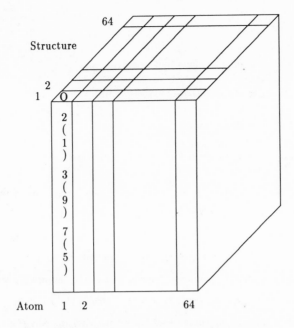

Figure 7.4: Vertical storage of structures.

occupies a row of array store. After reading the structures from the host disk unit, it is necessary to execute a data conversion routine. In this case, however, no such conversion took place; instead, a PDT was written which transformed the data during the remapping stage; this technique resulted in a reduced execution time. The physical mapping of structures, in the DAP array store, before and after remapping, is shown in Figures 7.3 and 7.4 respectively.

7.7.2 Compression Stage

The area of array store containing the structure data can be regarded as a sequence of logical data, rather than ASCII characters, by using the *EQUIVALENCE* statement of Fortran 77. Access to this logical data is gained by processing the structure data in a plane-wise sequence; that is, a single bit from all of the PE's is processed simultaneously. The data in a single line of a connection table may be divided into a number of fields: atom-type, neighbour label, bond-type. The processing of a field

135

involves several stages:

1. Select ASCII characters corresponding to a field.

2. Transform the data to a compressed representation.

3. *EQUIVALENCE* to binary code.

4. Extract binary data and copy to compressed workspace.

For example, consider the processing of the atom-type field, consisting of two characters. If the atom label is Pb (lead), then this is transformed by a table look-up procedure to the atomic number of lead, 82. This number is stored as an eight bit integer. The seven least significant bits are then extracted and copied to the compressed workspace. Housekeeping information required in the compressed structure records, such as the neighbour connectivity, may be calculated during the compression stage and then copied to the compressed workspace after the ASCII record has been processed.

All such operations are completed in parallel for 64 structures. The processing is efficient since the data is aligned in all the PE's and maximum synchronisation can therefore be achieved. After compression, each connection table line is represented as a short bit-string residing under a PE in vertical mode.

7.7.3 Bit-String Concatenation

The bit-strings in an array store row represent a single chemical structure. These strings are of variable length and must be combined together to form a single bit-string representing the structure. For this stage, it is convenient to remap the data back to horizontal mode; a further PDT is used for this.

After remapping, each bit-string occupies a single plane, with all bit-strings starting at the same relative position. In order that bit-strings do not overwrite each other, it is necessary to realign the bit-strings. Consider the planes shown in Figure 7.5. Before merging, the second plane is shifted 40 places in long vector

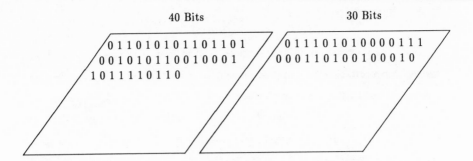

Figure 7.5: Bit-string planes before concatenation.

mode to provide space for the first plane. The two planes can then be merged by forming a new plane which is the logical OR of the original planes. The process is repeated for all planes comprising a structure, and for every structure. The result is 64 concatenated bit-strings.

7.7.4   Bit-String Record Size

For efficient processing of the binary structure file, it is necessary for each structure record to occupy a fixed amount of storage. This approach permits a pre-calculated number of structures to be read from disk in a single operation, thus minimising the disk latency associated with a sequence of read operations. PDT's work most efficiently with data for which the bit representation is a power of two; it was therefore decided that each structure should occupy the next power of two number of bits greater than its size. For ease of implementation, this number was chosen to be 4Kbits (a single plane of array store) for each structure.

The compression procedure is repeated for each 64-structure block of the input file until the whole file has been compressed. In an operational chemical information system, all of the day's new structures could be added to the structure file by a batch run of this compression program in the evening.

137

7.8 Run-Time System

At run-time, the compressed data must be read in from backing store and transformed into a suitable representation for comparison with a query. For the Ullmann algorithm, bond adjacency matrices are required together with a matrix defining tentative mappings between atoms (the B and M matrices respectively, see Section 6.3.1). Assuming structures are to be searched in blocks of 4,096, the pre-processing stage involves the generation of 16,384 DAP-sized logical matrices (each bond requires three bits, and when the M matrix is included, each structure generates four sets of 4,096 planes). This data must be generated at search time; and it is therefore important that fast algorithms can be found if adequate response times are to be achieved.

7.8.1 Data Mapping

Structures are read from disk in blocks of 4,096 and stored in horizontal mode. For processing, it is necessary to distribute the structures on a one per PE basis. The data is remapped to vertical mode as illustrated in Figure 7.6.

7.8.2 Bond Matrix Generation

The bond adjacency matrix is a matrix whose elements specify bond types between connected atoms, with null values being included where no bond exists. The challenge is to process concurrently all of the bit-strings to generate such bond matrices efficiently. In this case, each bit-string is of variable length because:

1. each structure contains a variable number of atoms;

2. the connectivity of each atom varies.

It is therefore impossible to achieve complete synchronisation when processing many structures together.

138

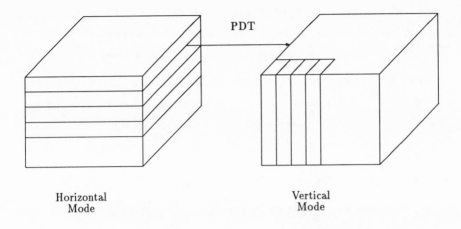

Horizontal
Mode

Vertical
Mode

Figure 7.6: Data remapping using a PDT.

As shown in Table 7.1, there are five distinct field types within a bit-string record. Each bit-string is processed sequentially starting with the *atom count* field. As the bit-string is processed, a flag records the field type of the data at the current position in the string. For each field type, a different code section is entered, and masking is used to make sure that the correct code is used to process the data according to a PE's flag. Since only four distinct fields occur regularly in a bit-string (the *atom count* field occurs only once in a structure), a high degree of parallelism is achieved.

The bond matrix is represented as a serial bit-string in the local memory of each PE. The data structure that this string represents is a $N * N * K$ array, where N is the maximum structure size and K is the number of bits used to represent the bond; here, $K = 3$. Hashing can be used to update the bond matrix. A pointer is calculated, given an atom number and a neighbour number:

$$pointer = (atom\ number - 1) * 3 + neighbour\ number$$

139

The bond-type corresponding to the atom and neighbour is then copied to the specified position in the array store. When many pointers are calculated together, the hash function returns a disparate set of pointer values. Since PE's cannot index memory on a local basis, the set of planes between the positions $pointer_{min}$ and $pointer_{max}$ must be scanned and updated, this being an unavoidable inefficiency.

After processing, each structure will be represented by a lower-triangular bond matrix (each bond is stored once only in the bit-string). The full matrix may be calculated by reflecting the information into the upper-triangular part of the matrix to form a symmetric bond matrix. Again, this processing is performed in parallel for all structures.

7.8.3 Generating Tentative Mappings

The M matrix specifies tentative mappings between each query atom and all the database structure atoms. The condition used in this case is based on matching the augmented atoms of the two structures. That is, the atom types and also the pattern of surrounding atoms and bonds of the query atom must be a subset of those in the database structure atom. To compute this condition for the set of atom pairs, it is necessary to set up a temporary workspace for the query and database structures. This workspace is a 2-D array where entry i, j is a tuple containing the atomic number of atom j and the bond-type connecting atom i to atom j. Entries corresponding to atom pairs which are not connected have zero tuples.

Consider the workspace generated for the structure in Section 7.5.1:

	1	2	3	4	5
1	0	6,1	0	0	6,1
2	8,1	0	6,2	0	0
3	0	6,2	0	6,1	0
4	0	0	6,1	0	6,2
5	8,1	0	0	6,2	0

The generation of tentative mappings involves a comparison of the query and database structure workspaces. For efficiency the tuples in each row are sorted, the major key is the atomic number and the minor key is the bond-type:

	1	2	3	4	5
1	6,1	6,1	0	0	0
2	8,1	6,2	0	0	0
3	6,2	6,1	0	0	0
4	6,2	6,1	0	0	0
5	8,1	6,2	0	0	0

Each entry in the correspondence matrix $M(i, j)$ can then be calculated by comparing row i of the query workspace with row j of the database workspace (assuming that the atomic numbers match).

Row Matching

A pointer in each row points to the next tuple to be processed, and the pointers move along each row sequentially. The algorithm for row matching is:

TYPE $tupleptr, ptr$ {tuple pointer}
VAR $qptr, dptr$: $tupleptr$ {pointers for query and database rows respectively}
FUNCTION $end(ptr : tupleptr)$: BOOLEAN
{returns TRUE if ptr points to null data, FALSE otherwise}
FUNCTION $match(ptr1, ptr2 : tupleptr)$: BOOLEAN
{returns TRUE if the tuple pointed to by $ptr1$ matches the tuple pointed to by $ptr2$}
FUNCTION $next(ptr : tupleptr)$: $tupleptr$
{returns the next tuple after the tuple pointed to by ptr}

WHILE NOT($end(qptr)$ OR $end(dptr)$) DO
BEGIN
   WHILE not($match(qptr, dptr)$) DO
     $dptr := next(dptr)$;

$qptr := next(qptr)$

END;

$match := \text{NOT}(end(dptr))$.

Implementation

The workspace data structure may be generated directly given a list of atomic numbers, their associated labels and the bond adjacency matrix; this operation is performed in parallel in each PE. The next stage is to sort the rows of the workspace; a parallel bubble-sort algorithm is used for this purpose since the number of records to be sorted is at most 16. Code was written which was optimised for the record lengths involved, in this case 10 bits.

Each M matrix is of size $P_\alpha * P_\beta$ and each element $M(i, j)$ is generated in parallel for all structures by comparing row i from the query with row j from the database structure. The plane generated is then copied to the $i * j$th plane of the array store containing M. This process is repeated serially for all values of i, j.

After processing, each PE has the bond adjacency and tentative mapping matrices associated with a single structure in local memory. For algorithm II, the Ullmann search can commence immediately; for algorithm I, the data is remapped to horizontal mode.

7.8.4 System Performance

The approximate time taken to read 4,096 molecules from disk was 15 seconds. This time includes the generation of tentative mappings, based on augmented-atom substructural equivalence, between each query atom and every database atom. It is believed that further optimisation of the code might reduce this time by a factor of four. The reader should note that the execution time quoted is for the worst-case scenario, i.e., for files consisting entirely of large molecules; this is a result of the data-parallel approach used in the design of the system.

142

7.9 Future System Design

At present the substructure searching system does not include a bit-string search; however, the system could readily incorporate such a stage. Conventionally, bit-strings are stored in a separate file, which is read during the screening stage and a record made of all those structures passing the bit-string comparison match. These structures are then retrieved from the structure file for a detailed atom-by-atom comparison with the query. This process involves many disk reads since the structures passing the filter stage will be evenly distributed throughout the structure file.

The alternative considered here is to include the bit-strings with their associated structure bit-strings in a single file. Structures are read from disk in blocks of 4,096. For the screening search, the bit-strings are remapped to vertical mode with the structure data remaining in horizontal mode. The structure data associated with those bit-strings passing the screening stage is retained and all the other data discarded. Further blocks of 4,096 structures are processed in this way until 4,096 structures have been accumulated or the end of the file is reached. At this point, the subgraph isomorphism procedure is invoked. Although this method involves reading the whole structure file, it is efficient because:

1. Structures occupy a small amount of storage in the compressed form, typically less than the bit-string.

2. Disk latency is minimised since the number of disk read operations is low.

The block size need not be 4,096; if enough array store were available, an arbitrary number of structures could be input in a single read operation. Each structure occupies about 2 Kbits, and 4,096 structures require 1 Mbyte of storage. Given a 128 MByte machine (assuming 64 Mbytes is required for the algorithm) 262,144 structures could be processed per disk access.

Chapter 8

Macromolecule Ranking on the DAP

8.1 Introduction

The previous chapters have been concerned with the processing of small, 2-D chemical structures. In this chapter a different application of the DAP to the processing of chemical structures is presented: that of calculating the similarity between 3-D protein structures.

8.2 Background

A protein is a linear polymer: a chain of sub-units linked in a continuous sequence. The sub-unit of a protein is an amino acid: there are twenty naturally occurring examples. Proteins may be described at different levels of detail; at the primary level the amino acid sequence is recorded, using an alphabet of twenty symbols, one for each amino acid. Sequences of amino acids form into helices and chains and these characterise the secondary structure of a protein. Finally, the tertiary structure of a protein is the complete three-dimensional conformation of the amino acids. In this chapter we will only be concerned with the secondary level of description.

144

The secondary structure of a protein is formally defined by Kendrew *et al.* [95] as 'the spatial arrangement of its main-chain atoms without regard to the conformation of its side chains or to its relationship with other segments'. With the determination of the structures of a large number of proteins, it has become evident that there are commonly occurring structural motifs, often definable at the level of secondary structure.

The last few years have seen the widespread adoption of database systems for the handling of information about 3-D macromolecular species, using co-ordinate data obtained from X-ray crystallography and NMR (Nuclear Magnetic Resonance), in addition to providing facilities for graphical analysis and display of protein structures [40,80]. An example of this is the work at Sheffield University, which has resulted in a range of searching techniques for the protein structures in the Protein Data Bank [26], the primary source of 3-D data for macromolecules. The techniques involve the representation and searching of proteins at the secondary structure level [13] and include a novel system, called POSSUM, that allows substructure searches to be carried out for user-defined secondary structure motifs [95].

8.3   The POSSUM System

POSSUM (Protein Online Substructure Searching - Ullmann Method) [95] is a retrieval system that allows the user to specify a 3-D query motif composed of helices and/or strands and to search for all occurrences of this motif in the protein structures of the Protein Data Bank.

8.3.1   Representation of Secondary Structure Elements

The structure representation used in POSSUM is based on the fact that the common helix and strand secondary structure elements, or SSE's, are approximately linear, repeating structures and that an SSE can hence be described by a vector along its linear axis [7]. The set of vectors corresponding to the secondary structure elements in a protein can then be used to describe the structure of that protein in 3-D space, with the relative orientation of the helix and strand elements being defined by the inter-line angles and distances. The topography of a query motif can be described in the same way, and the retrieval of motifs consisting of helix and/or strand elements can thus be achieved by a matching operation based on these inter-line angles and distances.

The protein and query motifs may be represented as labelled graphs, with the nodes of the graphs corresponding to the linear representation of the helices and strands, and the edges to the inter-line angles or distances. For small molecules, graphs provide a much lower level of description with nodes representing atoms and edges corresponding to bonds, in the case of 2-D structures, or inter-atomic distances, in the case of 3-D structures [90] (as discussed in previous chapters). For substructure searching of proteins this level of detail is unnecessary and impractical, since a protein may contain thousands of atoms.

A graph representation of macromolecules allows the use of subgraph isomorphism algorithms for the detection of motifs in protein structures. In this case the subgraph isomorphism algorithm proposed by Ullmann [124] was used as it was found to be efficient for 3-D substructure detection in small molecules in earlier studies [28]. A full description of the algorithm can be found in Chapter 6.

Protein data bank structures are represented for search by matrices containing inter-axial angles and also midpoint and closest approach distances. The availability of these three types of data provides a range of searching options depending upon the requirements of a particular query, for example:

- By setting the distance tolerance to a very large figure, eg, $\pm 500 \mathring{A}$, the user can search for the pattern using purely angular constraints, although this does not generally give a very precise search.

- An angular tolerance of $\pm 360°$ will allow the user to search for the pattern using only distance constraints.

- The user may also specify distance tolerances for the distance between the midpoints of the axis lines and also the closest approach distance.

Searches using a range of typical query motifs demonstrated the effectiveness of the program for detecting both known and previously unknown occurrences of well known structural motifs in the Protein Data Bank and the program is now being used on a routine basis in the Department of Molecular Biology and Biotechnology at the University of Sheffield [13,95]. A more detailed description of POSSUM is given by Mitchell [94].

8.3.2 Calculation of Inter-Motif Similarity

POSSUM is an example of a partial match retrieval system, i.e., one where the retrieved structures contain the query pattern in addition to other substructural features. Best match or nearest neighbour searching systems allow the retrieval of structures that are related in some quantitive way to an input target structure. Such techniques have been applied to the processing of 2-D chemical molecules for applications such as structure-property correlation and the selection of compounds for biological screening [132]. In this section the best match search algorithm which is used to rank output from a POSSUM search is described.

When POSSUM was first designed the intention was that searching should be based solely upon the inter-line angles. However, initial results demonstrated clearly that both angular and distance information need to be specified in a query if an effective search is to be carried out. There are, however, two limitations associated with the use of the distance information.

- The precise form of the distance constraints is dependent on the type of query which is to be searched for.

- No explicit account is taken of the size of the SSE's, i.e., the lengths of the lines in 3-D space. It is thus possible to obtain hits which, while satisfying the information in the query statement, differ considerably from the actual motif represented by this statement.

147

In POSSUM, the best match search algorithm seeks to overcome these two limitations by means of a post-processing step. The output from a broadly defined POSSUM search is ranked in order of similarity with the query, with those at the top of the ranking being most similar.

The overall topography of a motif is approximated by constructing a distribution of inter-line distances between component SSE's. Specifically, each line is divided into a series of equidistant points. Distances are calculated between all pairs of points on separate lines, whereas the basic substructure search algorithm in POSSUM utilises only a single distance (either the mid-point or closest approach distance) to characterise the structural relationship between each pair of lines. The set of all distances is used to build a frequency distribution, which characterises the motif. The degree of similarity between a query motif and a database motif is then measured by the extent of the agreement between the two distance distributions. This procedure takes into account the lengths of the lines representing SSE's and this will be evident in the ranked output produced by the similarity algorithm.

The algorithm involves the following steps:

1. Each vector composing a motif is split into segments of equal lengths so as to obtain a number of points. The length of the segments is T; the smaller the value of T, the greater the number of points there are along the lines.

2. A frequency distribution is created containing a number of inter-point distance range categories. A parameter R defines the range of distances to be included in a single category. Thus, if an R value of 1.0 is specified, then the distance range categories will be:

$$0.0 \longrightarrow 0.9$$
$$1.0 \longrightarrow 1.9$$
$$2.0 \longrightarrow 2.9 \text{ etc.}$$

148

3. The distance between each point on every line and every other point on every other line is calculated. Each distance is used to update the frequency distribution associated with the motif.

4. Steps 1-3 are carried out for the query and all database motif hits produced by the initial search.

5. The query distribution is compared with each of the database motif distributions. The similarity can be calculated using a statistical measure such as χ^2 or the sum of squared differences.

8.4   Experimental Details

Three sets of data were chosen to compare the performance of the serial and parallel algorithms; further details of these can be found in Grindley [62]:

- Calcium-Binding Fold(TNC2): this is a pattern of 2 β-strands and 3 α-helices which represents the calcium-binding site found within the muscle protein Toponin-C (Protein Data Bank code 2TNC).

- NAD-Binding Fold(NADX): this pattern represents part of the nucleotide-binding domain found within the protein lactate dehydrogenase (Protein Data Bank code 4LDH) and has 2 α-helices and 5 β-strands.

- 8-Stranded β-Barrel(BBAR): this is a pattern of considerable structural importance, the query being derived from the motif's occurrence in triose phosphate isomerase (Protein Data Bank code 1TIM).

The ability to specify values for the parameters T and R means that a wide range of experimental conditions is available for investigation. An increase in T results in a decrease in the number of points in each linear SSE and a corresponding decrease in the number of inter-point distances that must be evaluated. An increase in R results in a reduction in the number of categories included in the statistical calculation.

Mitchell [94] gives a detailed description of POSSUM and discusses its performance. The similarity algorithm and an analysis of its effectiveness when used with three commonly occurring queries is given by Grindley [62]. In addition, the similarity measure that is used will also affect the rankings; in this case, a single statistical measure, the sum of squared differences, was used because the choice of measure did not affect the efficiency of the algorithm. In each case, the following procedure was adopted:

1. The initial substructure search was executed with tolerance values set such that a large number of hits was produced.

2. Rankings of these hits were generated using a range of T and R values: $0.10 \leq T \leq 10$ and $0.5 \leq R \leq 5.0$. In previous versions of the program the lowest value of T used was 0.5; for this value the run-time was of the order of hours of CPU time on a VAX mini-computer [62]. This can be explained by the requirement to process a large number of points. It was hoped that increased efficiency would permit the parallel algorithm to accommodate small values of T.

8.5 Serial Implementation of the Algorithm

The similarity algorithm has been shown to produce highly effective rankings; it is however, extremely time-consuming in execution [62]. An inspection of the pseudo code of the algorithm explains why the algorithm is so computationally demanding.

Assuming:

$NLINES$ = the number of vectors representing SSE's in the motif,

$NPOINTS(I)$ = the number of points on vector $LINE(I)$,

$FREQDIST$ = the frequency distribution of the database motif, and,

$QFREQDIST$ = the frequency distribution of the query motif,

the algorithm is:

FOR each database motif resulting from the POSSUM search DO

150

BEGIN

   set zero all the elements of the integer array $FREQDIST$;

   FOR $I := 1$ TO $NLINES - 1$ DO

    FOR $J := I + 1$ TO $NLINES$ DO

     FOR $K := 1$ TO $NPOINTS(I)$ DO

      FOR $L := 1$ TO $NPOINTS(J)$ DO

      BEGIN

        calculate distance between the points $LINE(I, K)$

        and $LINE(J, L)$;

        increment the element of $FREQDIST$ corresponding to

        this distance by one

      END;

   calculate the degree of similarity between $FREQDIST$ and $QFREQDIST$

END;

display the database motifs in order of decreasing similarity.

Consider a single motif. First, each line in the motif is split into a number of segments: this process is clearly linear in complexity and this is also true for the frequency distribution creation stage. The crucial part of the algorithm is the distance calculation stage. The total number of distance calculations involved is given by:

$$\sum_{I=1}^{NLINES-1} \sum_{J=I+1}^{NLINES} NPOINTS(I) * NPOINTS(J) \qquad (8.1)$$

If the total number of points, N, is given by:

$$N = \sum_{I=1}^{NLINES} NPOINTS(I)$$

then the complexity of Equation 8.1 is $O(N^2)$. When T is small, N will be large since $N \propto \frac{1}{T}$, and this corresponds to large numbers of points resulting from a small segment length. Direct addressing achieved by hashing a distance value may be used to update the frequency distribution efficiently, and the complexity of this stage is negligible. The comparison of two frequency distributions is of linear complexity.

151

In terms of complexity, the dominant part of the algorithm is the distance calculation stage, since nearly N^2 Euclidean distances, in 3 dimensions, must be calculated. Unless T is set to a large value, many distance calculations are required and searching can be extremely slow given the MicroVAX-II hardware on which POSSUM was originally implemented. Clearly each distance calculation is independent and several may be calculated together given the appropriate hardware. The problem would thus seem to be ideally suited to a highly parallel SIMD type machine, since the same operation is applied repeatedly to a large set of homogeneous data. A parallel version of the distance calculation stage was prepared for implementation on the DAP.

8.6 Parallel Ranking Algorithm I

The parallel algorithm involves distributing the co-ordinates of points among the PE's of the DAP array, one point per PE, in long vector mode. An individual point is then selected and its co-ordinates broadcast from the MCU, thus allowing the parallel calculation of the distances between this point and all other points in the motif. Associated with each PE is a frequency distribution, this is updated after each distance calculation. If updates are to be carried out in parallel, direct indexing by hashing may not be used as in the serial case. This is because there is no hardware facility to allow independent indexing within an individual PE. Therefore, indexing must be performed sequentially in order to update the frequency distributions within all PE's simultaneously. If there are S partitions in the frequency distribution, then the frequency distribution update stage will require S serial steps. The pseudo code for the parallel ranking algorithm is as follows:

FOR each database motif resulting from the POSSUM search DO
BEGIN
    set to zero all the elements of an integer array $FREQDIST$;
    FOR $I := 1$ TO N
    BEGIN

broadcast co-ordinates of $POINT(I)$ from MCU to the array;

calculate all distances in parallel;

FOR $J := 1$ TO S DO

for all PE's in parallel:

IF distance $\in RANGE(J)$ THEN $FREQDIST(J) := FREQDIST(J) + 1$

END;

calculate degree of similarity between $FREQDIST$ and $QFREQDIST$

END;

display the database motifs in order of decreasing similarity.

8.6.1 Complexity of the Algorithm

Consider the calculation of the frequency distribution for a single motif. The above description assumes that the number of points N is less than or equal to the number of DAP PE's, or $N \leq ES^2$, where ES is the edge size of the DAP. If $N > ES^2$, then the problem must be broken into $\geq \frac{N}{ES^2}$ fragments (this type of problem decomposition is known as a sheet mapping). For each fragment/sheet, N co-ordinates must be broadcast, so we have,

$$\frac{N * N}{ES^2} = \frac{N^2}{ES^2} \text{ operations.}$$

Therefore, the complexity for an $O(ES^2)$ array processor is, $O(\frac{N^2}{ES^2})$, and thus we have:

$$\text{complexity} = \begin{cases} O(N) & \text{if } ES^2 \geq N \\ O(N^2) & \text{otherwise.} \end{cases}$$

Therefore, the parallel distance algorithm is linear in complexity when the number of points is not greater than the number of DAP PE's; otherwise, the complexity is the same as for the serial version. However, the coefficient term $\frac{1}{ES^2}$ ensures that the parallel version will have to perform less operations than the serial.

The frequency distribution update stage must also be considered as this is per-

153

formed after every distance calculation. The update is a sequential search involving S steps; taking this into consideration the complexity is $O(\frac{N^2 S}{ES^2})$. In most cases, $ES^2 > N$ for a motif, so finally we have

$$
\text{complexity} = \begin{cases} O(NS) & \text{for an } O(ES^2) \text{ array processor} \\ O(N^2) & \text{for a serial processor.} \end{cases}
$$

Because S only influences the parallel algorithm, it is clear that the choice of the R parameter will affect the speed-up.

8.6.2 Results

Run times for the three queries are shown in Tables 8.1, 8.2 and 8.3; $T(S)$ and $T(P)$ are the serial and parallel execution times respectively and $S(P)$ is the speedup or performance ratio. The results coincide with the complexity analysis: for the serial version, $\frac{1}{T}$ varies with the square of the run-time, and the longest run-times are recorded for small values of T. For the parallel algorithm the relationship between $\frac{1}{T}$ and the run-time is linear for a fixed value of R. When T is constant, the results show that the correspondence between $\frac{1}{R}$ and the run-time is also linear. In the serial case, the R parameter has no effect on execution time.

The best speed-ups are achieved for large structures, since more points will require processing (this corresponds to a smaller T value). In this case, the largest structure was BBAR, followed by NADX and TNC2. The magnitude of the speed-up varies in the range 0.1...8.8, with the best results obtained for small segment lengths. This indicates that the DAP would be best suited for making detailed rankings of motifs.

The absolute magnitude of the speed-ups might seem low for such a highly parallel application. This can be explained by the type of processing involved, which is a mixture of integer and real arithmetic. Serial processors, such as the IBM 3083, have hardware optimised for such processing, whereas, in the DAP's case, non-Boolean operations are constructed in software from primitive logical operations.

8.7 Parallel Ranking Algorithm II

In order to make more efficient use of the DAP, the ability to process several motifs together was required. This was because most motifs were not large enough to fill a DAP array (unless extremely small values of T were used). Thus only a fraction of PE's were utilised since those PE's without motif data performed no useful work. Optimal efficiency can be achieved by assuming a separate co-ordinate system for the points of each motif. A point from a motif is selected and the co-ordinate systems of all other motifs are shifted so that a single point from each motif occupies the same position as the original point. The distance between this point and all other points in each motif may be calculated in parallel. This procedure is useful since it is the distances, not the actual co-ordinates, that are of interest.

By first selecting the motif with the largest number of points and repeating the procedure for every point in the motif, all inter-point distances, in every motif, may be calculated efficiently. The points from many motifs can be used in order to occupy as many DAP PE's as possible. Therefore, for a given set of motifs, the most efficient use of the DAP is made. The only overhead is the co-ordinate system shift which is made before calculating distances. The pseudo code for the algorithm is:

Select the motif with the largest number of points,
let this be motif M with k points;
DO P=1 TO k
    shift the co-ordinate systems of all motifs so that the
    Pth point in each shares the same position;
    calculate the distance between the Pth point in motif M
    and all the other points in every motif;
    update the frequency distributions for every motif.

Alternatively, a more rigorous description may be given.

The set of co-ordinates in one dimension are:

$$\{x_{ij} \mid i = 1, \ldots, n \wedge j = 1, \ldots, points(i)\}$$

where $points(i) = $ the number of points in motif i

and $n = $ the number of motifs

$k = max(points(i))\ i = 1, \ldots, n$

DO $p = 1, \ldots, (k-1)$

$\quad \forall i$ such that $points(i) \geq p$

$\quad \Delta_i = (x_{kp} - x_{ip})$

$\quad \forall j \neq p$

$\quad\quad x_{ij}^{(2)} = x_{ij} + \Delta_{ij}$

$\quad\quad dist(j \text{ to } p)_i = x_{kp} - x_{ij}^{(2)}$

Example:

Consider two sets of points in one dimension:

$$A = \{16, 47, 109\}, B = \{36, 134, 88, 72\}.$$

Select the set with the largest number of points, in this case set B, and take the first point from the set, b_1. We now shift the set A so that its first point, a_1, occupies the same co-ordinate as b_1. This is done by adding the appropriate scalar, Δ, to each point in A, $\Delta = b_1 - a_1 = 36 - 16 = 20$. We thus now have:

$$A = \{36, 67, 129\}.$$

The distances between the first point, and all other points, for each set, may now be calculated by subtracting b_1 from the members of the sets A and B. Finally we have:

$$distance_{a_1} = \{0, 31, 93\}, distance_{b_1} = \{0, 98, 52, 36\}$$

(where the set $distance_{j_k}$ is the set of distances from element k of set j to all other members of the set).

8.7.1 Results

Run times for the three queries are shown in Tables 8.4, 8.5 and 8.6. As before, $T(S)$ and $T(P)$ are the serial and parallel execution times respectively and $S(P)$ is the speedup or performance ratio. The results for this algorithm follow the same pattern as the first. The times are at least as fast as the first algorithm in all cases. The most striking improvements are seen when large values of T are used; when it is possible to store, and to process, the points of several motifs stored in the same DAP-sized matrix. Efficiency is still maintained by calculating inter-point distances for all motifs simultaneously: the first algorithm treated each motif separately. If T is sufficiently small to generate exactly one DAP matrix of points, the performance of the two algorithms will be similar. In general, this algorithm will be more efficient than the first whenever data from more than one motif is stored in the same DAP-sized matrix.

8.8 Bit-Level Algorithm

The aim of this algorithm   is to eliminate the need for any integer or real arithmetic by modelling the vectors which represent SSE's in the memory of the DAP. To achieve this,continuous space is made discrete by dividing a subset of $\mathbb{R}^3$ (three-dimensional real space) into equal subspaces. With each subspace there is an associated Boolean variable. A point on a vector which falls within a given subspace has its associated variable set to TRUE; vectors may be plotted in this way. The city block distance measure is employed since a Euclidean distance calculation requires real arithmetic.

8.8.1 Algorithm

Assume an abstract data type called $SPACE$, which is a 3-dimensional Boolean array; then, the following functions are defined on $SPACE$:

157

$a, b, c, motif, motifcopy, point, neighbours, hits : SPACE;$

$s, count$:INTEGER;

z :BOOLEAN;

- Logical AND: The result of the operation $AND(a, b)$ is a space where each element is the logical AND of similar elements in a and b.

- Logical NOT: The space $NOT(a)$ consists of a logical negation of the elements of a.

- MODULAS: If $s = MODULAS(a)$, then s is the sum of all the elements of a which are TRUE.

- PROPAGATE: If $a = PROPAGATE(b)$, each TRUE element of b is negated and then its nearest neighbours along each axis are set to TRUE, forming the space a.

- ZEROSPACE(a): If $z = ZEROSPACE(a)$, then z is TRUE if there are no TRUE elements in the space a, otherwise z is set to FALSE.

- FIRST(a): If $b = FIRST(a)$, then b is a copy of a with every element set to FALSE except the first.

Given these functions, the algorithm is as follows:

For each point in the motif DO
BEGIN
   $motif := motifcopy;$
   $point := FIRST(motif);$
   $count := 0;$
   REPEAT
     $count := count + 1;$
     $neighbours := PROPAGATE(point);$
     $hits := AND(motif, neighbours);$

$$freqdist := freqdist + MODULAS(hits);$$
$$motif := AND(motif, NOT(hits));$$
UNTIL $ZEROSPACE(motif)$

END.

The algorithm utilises the same principles as a radar to find the distances from a point to all other points. Bits are propagated outwards in all directions (north, east, south, up and down) from a point, in steps of one subspace. After each propagation, a check is made of how many propagating bits have collided with points in the motif. The distance between the original point and the motif point is given by the number of propagations. Thus, after each propagation, the frequency distribution may be updated. The propagation finishes when every motif bit is reached; the process is then repeated for the next point in the motif.

While it would obviously be desirable for propagation to take place from all points in the motif simultaneously, this is not possible in practice. If two or more propagating bits occupy the same subspace a greater precision than one bit is required to record their presence, unless information is to be lost. After one propagation, it is possible that six bits might occupy a subspace, after two propagations, 36 etc., and the algorithm is no longer a simple Boolean one. In order to maintain bit-level precision, it is therefore necessary to treat each point in the motif individually.

8.8.2 Implementation

The data type $SPACE$ is represented on the DAP as an array of ES logical matrices. Thus, each plane of a space is processed efficiently, but since only one plane can be processed at a time, all operations involving a $SPACE$ include a serial step. Restricting the size of a $SPACE$ to DAP dimensions introduces limitations on the values of T and R. If a very small T is specified, then many points will fall within the same subspace and will be treated as a single point. To allow smaller values of T to be used, a bigger $SPACE$ is required and, if efficiency is to be maintained, a parallel computer with more PE's. In our work, the size of $SPACE$ was fixed at ES^2 and the optimal value for T was used,given by dividing the co-ordinate system range, required to represent a motif, by the number of subspaces, ES:

$$T = \frac{\text{co-ordinate range}}{ES} = \frac{400}{64} = 6.25 \text{ Å}.$$

The city block distances generated by the algorithm are in terms of the number of subspaces of size T between two points, resulting in a fixed $R = T$. This is because distances cannot be measured with a greater precision than the subspace size. The frequency distribution contains $3ES$ partitions, $3ES$ being the maximum city block distance between any two points in a cube of side length ES.

8.8.3 Results

Times are given for each query motif in Table 8.7. These are slow compared to the previous algorithms with similar values for R and T. This can be explained by the algorithm being too computationally intensive for an ES^2 DAP. In order to run the algorithm efficiently, it is necessary to map the data for a motif directly onto the DAP processor array, which would require a ES^3 DAP. Operations on the data type $SPACE$ would then be carried out very rapidly: in this case, it is estimated that the algorithm would be 1-2 orders of magnitude faster than the current bit-level implementation.

8.9 Conclusions

Of the three parallel algorithms considered, for implementing the ranking stage of a POSSUM search, the first two were similar and relied on calculating many distances simultaneously to achieve a high throughput. The first algorithm was improved by a co-ordinate shift which allowed several motifs to be processed together. Both demonstrated that substantial reductions in execution time can be achieved when using the DAP; however, the exact speed-up was crucially dependent on the

160

input parameters T and R. This was because of the lack of a hardware indexing facility within DAP PE's.

The third parallel algorithm was radically different from the others and relied on modelling motifs in the DAP array store. An iterative procedure was then applied to detect distances between locations in the array store by bit propagation. It was found that this algorithm, although elegant, required a machine with more PE's than the currently available DAP to achieve the flexibility and efficiency of the previous two algorithms. The third algorithm undoubtedly has good potential.

$R\backslash\text{Å}$	$T\backslash\text{Å}$	$T(S)$	$T(P)$	$S(P)$
	0.10	97.42	125.19	0.78
	0.25	22.23	58.39	0.38
	0.50	6.25	29.63	0.21
0.50	0.75	3.16	20.33	0.16
	1.00	2.11	15.74	0.13
	5.00	0.69	4.76	0.14
	8.00	0.64	3.66	0.17
	10.00	0.63	0.63	1.00
	0.10	97.87	81.80	1.20
	0.25	22.36	37.80	0.59
	0.50	6.26	19.55	0.32
1.00	0.75	3.17	13.26	0.24
	1.00	2.09	10.37	0.20
	5.00	0.68	3.08	0.22
	8.00	0.63	2.36	0.27
	10.00	0.62	2.16	0.29
	0.10	97.78	55.29	1.76
	0.25	22.35	25.72	0.87
	0.50	6.24	13.12	0.48
2.50	0.75	3.16	9.11	0.35
	1.00	2.09	7.03	0.30
	5.00	0.68	2.02	0.34
	8.00	0.63	1.53	0.41
	10.00	0.62	1.42	0.44
	0.10	97.18	47.33	2.05
	0.25	22.32	21.80	1.02
	0.50	6.24	11.07	0.56
5.00	0.75	3.15	7.56	0.42
	1.00	2.09	5.85	0.36
	5.00	0.67	1.68	0.40
	8.00	0.63	1.28	0.49
	10.00	0.62	1.17	0.53

Table 8.1: Execution times for TNC2 query using algorithm I.

$R\backslash\text{Å}$	$T\backslash\text{Å}$	$T(S)$	$T(P)$	$S(P)$
	0.10	1170.85	*	
	0.25	204.28	67.99	3.00
	0.50	54.82	34.49	1.59
0.50	0.75	26.63	23.21	1.15
	1.00	16.88	17.63	0.96
	5.00	3.80	5.21	0.73
	8.00	3.33	3.92	0.85
	10.00	3.24	3.66	0.89
	0.10	1170.94	*	
	0.25	204.23	44.23	4.62
	0.50	54.84	22.42	2.45
1.00	0.75	26.63	15.20	1.75
	1.00	16.86	11.43	1.48
	5.00	3.83	3.32	1.15
	8.00	3.63	2.55	1.32
	10.00	3.32	2.35	1.37
	0.10	1167.35	*	
	0.25	203.80	30.24	6.74
	0.50	54.74	15.29	3.58
2.50	0.75	26.60	10.22	2.60
	1.00	16.82	7.87	2.14
	5.00	3.83	2.22	1.73
	8.00	3.42	1.67	2.05
	10.00	3.26	1.54	2.12
	0.10	1167.37	*	
	0.25	203.80	25.41	8.02
	0.50	54.74	12.93	4.23
5.00	0.75	26.59	8.63	3.08
	1.00	16.81	6.62	2.54
	5.00	3.82	1.88	2.03
	8.00	3.41	1.41	2.42
	10.00	3.29	1.28	2.57

Table 8.2: Execution times for NADX query using algorithm I.
* DAP array store limit exceeded for these parameters.

$R\backslash\text{Å}$	$T\backslash\text{Å}$	$T(S)$	$T(P)$	$S(P)$
	0.10	1304.86	*	
	0.25	225.06	68.35	3.29
	0.50	60.26	34.48	1.75
0.50	0.75	28.97	24.14	1.20
	1.00	17.95	18.18	0.99
	5.00	3.49	5.35	0.65
	8.00	2.98	4.01	0.74
	10.00	2.91	3.72	0.78
	0.10	1303.65	*	
	0.25	228.48	45.02	5.08
	0.50	60.53	22.70	2.67
1.00	0.75	17.95	15.93	1.50
	1.00	17.95	11.95	1.50
	5.00	3.49	3.41	1.02
	8.00	2.95	2.58	1.14
	10.00	2.91	2.39	1.22
	0.10	1302.94	*	
	0.25	228.56	30.33	7.54
	0.50	60.52	15.50	3.90
2.50	0.75	28.93	10.72	2.70
	1.00	17.93	8.08	2.22
	5.00	3.47	2.29	1.52
	8.00	2.94	1.72	1.71
	10.00	2.88	1.61	1.79
	0.10	1301.16	*	
	0.25	228.64	25.80	8.86
	0.50	60.53	12.99	4.66
5.00	0.75	28.93	9.15	3.16
	1.00	17.91	6.85	2.61
	5.00	3.45	1.93	1.79
	8.00	2.94	1.45	2.03
	10.00	2.87	1.32	2.17

Table 8.3: Execution times for BBAR query using algorithm I.
* DAP array store limit exceeded for these parameters.

164

$R\backslash\text{Å}$	$T\backslash\text{Å}$	$T(S)$	$T(P)$	$S(P)$
	0.10	97.42	24.74	3.94
	0.25	22.23	7.88	2.82
	0.50	6.25	3.00	2.08
0.50	0.75	3.16	1.91	1.65
	1.00	2.11	1.59	1.33
	5.00	0.69	0.72	0.96
	8.00	0.64	0.65	0.98
	10.00	0.63	0.63	1.00
	0.10	97.87	16.44	5.95
	0.25	22.36	5.32	4.20
	0.50	6.26	2.08	3.01
1.00	0.75	3.17	1.32	2.40
	1.00	2.09	1.06	1.97
	5.00	0.68	0.43	1.58
	8.00	0.63	0.38	1.66
	10.00	0.62	0.37	1.68
	0.10	97.78	11.44	8.55
	0.25	22.35	3.82	5.85
	0.50	6.24	1.54	4.05
2.50	0.75	3.16	0.97	3.26
	1.00	2.09	0.76	2.75
	5.00	0.68	0.26	2.62
	8.00	0.63	0.22	2.86
	10.00	0.62	0.21	2.95
	0.10	97.18	9.69	10.03
	0.25	22.32	3.32	6.72
	0.50	6.24	1.34	4.66
5.00	0.75	3.15	0.85	3.71
	1.00	2.09	0.65	3.22
	5.00	0.67	0.20	3.35
	8.00	0.63	0.16	3.94
	10.00	0.62	0.15	4.13

Table 8.4: Execution times for TNC2 query using algorithm II.

$R\backslash\text{Å}$	$T\backslash\text{Å}$	$T(S)$	$T(P)$	$S(P)$
	0.10	1170.85	*	
	0.25	204.28	43.63	4.68
	0.50	54.82	14.35	3.82
0.50	0.75	26.63	8.60	3.10
	1.00	16.88	6.17	2.74
	5.00	3.80	1.24	3.06
	8.00	3.33	1.35	2.47
	10.00	3.24	0.69	4.70
	0.10	1170.94	*	
	0.25	204.23	23.40	8.73
	0.50	54.84	10.09	5.44
1.00	0.75	26.63	6.31	4.22
	1.00	16.86	4.54	3.71
	5.00	3.83	0.89	4.30
	8.00	3.36	0.95	3.54
	10.00	3.22	0.47	6.85
	0.10	1167.35	*	
	0.25	203.80	21.09	9.66
	0.50	54.74	7.63	7.17
2.50	0.75	26.60	4.89	5.44
	1.00	16.82	3.60	4.67
	5.00	3.83	0.70	5.74
	8.00	3.42	0.71	4.82
	10.00	3.26	0.34	9.59
	0.10	1167.37	*	
	0.25	203.80	18.07	11.28
	0.50	54.74	6.89	7.94
5.00	0.75	26.59	4.49	5.92
	1.00	16.81	3.32	5.06
	5.00	3.82	0.63	6.06
	8.00	3.41	0.63	5.51
	10.00	3.29	0.30	10.97

Table 8.5: Execution times for NADX query using algorithm II.
* DAP array store limit exceeded for these parameters.

$R\backslash$Å	$T\backslash$Å	$T(S)$	$T(P)$	$S(P)$
	0.10	1304.86	*	
	0.25	225.06	50.77	4.43
	0.50	60.26	16.91	3.56
0.50	0.75	28.97	10.93	2.65
	1.00	17.95	8.33	2.15
	5.00	3.49	3.78	0.92
	8.00	2.98	3.48	0.86
	10.00	2.91	3.40	0.86
	0.10	1303.65	*	
	0.25	228.48	35.12	6.51
	0.50	60.53	11.77	5.14
1.00	0.75	28.95	7.60	3.81
	1.00	17.95	5.68	3.16
	5.00	3.49	2.32	1.50
	8.00	2.95	2.05	1.44
	10.00	2.91	2.02	1.44
	0.10	1302.94	*	
	0.25	228.56	25.66	8.91
	0.50	60.52	8.78	6.89
2.50	0.75	28.93	5.63	5.14
	1.00	17.93	4.14	4.33
	5.00	3.47	1.43	2.43
	8.00	2.94	1.21	2.43
	10.00	2.88	1.17	2.46
	0.10	1301.16	*	
	0.25	228.64	22.58	10.13
	0.50	60.53	7.65	7.91
5.00	0.75	28.93	4.93	5.87
	1.00	17.91	3.66	4.89
	5.00	3.45	1.13	3.05
	8.00	2.94	0.92	3.20
	10.00	2.87	0.89	3.22

Table 8.6: Execution times for BBAR query using algorithm II.
* DAP array store limit exceeded for these parameters.

$R\backslash\text{Å}$	$T\backslash\text{Å}$	Query	$T(S)$	$T(P)$	$S(P)$
6.25	6.25	TNC2	0.65	58.99	0.01
6.25	6.25	BBAR	3.12	255.63	0.01
6.25	6.25	NADX	3.50	238.63	0.01

Table 8.7: Execution Times for bit-level Algorithm.

Chapter 9

Conclusions and Suggestions for Further Work

9.1 Conclusions

The DAP can provide a significant improvement in processing efficiency, when compared to a fast serial processor, for substructure searching of 2-D chemical structures. The calculation of similarity between protein structures can be performed very efficiently on the DAP, when compared with a large mainframe processor. However, the lack of local control within the DAP PE means that the precise level of speedup obtained is dependent on the input parameters.

A few general comments can be made about the DAP. Firstly, because of the bit-serial operation of the DAP PE it is often impossible to give a single figure for the level of speedup that may be obtained. Secondly, the lack of local control within a DAP PE places limitations on the type of algorithm that may be used. Often the DAP may have to perform considerably more work than a serial processor, and as a result algorithms which appear similar in construction may vary greatly in efficiency. Performance ratios of two or even three orders of magnitude have been reported for certain applications running on the DAP. Such applications are invariably highly data-parallel and utilise low precision arithmetic. Generally, problems are not so amenable to an efficient parallel implementation on the DAP. In order for the DAP to be suitable for a larger class of applications the PE architecture will need to be made more flexible. The introduction of the 'co-processor', which is a general purpose eight-bit processor attached to every PE, goes some way to achieving this objective. A further useful enhancement would be the introduction of memory addressing hardware into each PE. Such hardware would significantly increase the performance of the DAP for many database searching applications, including those reported in this book.

9.2   Suggestions for Further Work

It seems likely that the DAP could be made to achieve higher performance levels by fine-tuning existing code and the introduction of techniques to reduce PE inactivity. Also, the techniques developed could be applied to other problems in which subgraph isomorphism algorithms are used. There follows a list of suggestions for further work.

9.2.1 Increasing the Performance of Existing Code

1. Certain critical sections of code could be re-written to obtain the maximum possible performance. This might mean re-coding in APAL, or alternatively, re-structuring the software to take advantage of the performance benefits of the new Fortran-Plus compiler. It is believed that such a re-coding might yield a two-fold speedup in Algorithm I and up to a five-fold speedup in Algorithm II.

2. A new piece of hardware known as the 'co-processor' has been developed by AMT. The co-processor is an 8-bit processor, which is attached to each PE, specifically to increase the performance of multi-bit arithmetic. This hardware development should allow the relaxation procedure to be performed more efficiently for both parallel substructure algorithms.

9.2.2 Implementation of Algorithm IV

The so-called drip-feed algorithm reduces the effects of poor synchronisation in Algorithm II, by distributing new molecules to the PE array as PE's become idle. The implementation would involve writing run-time code to load-balance the PE array; such code could include routines from the PDT library. To fully test the code, however, a file in excess of 40,000 molecules would be required. Such a system would only be useful for large databases.

9.2.3 Implementation of Screening Algorithms

A screening stage could be incorporated into the DAP-based substructure search system reported in Chapter 7. It is thought that the performance of the parallel substructure search algorithms would be superior on a screened file since it has been shown that the DAP processes structurally similar molecules relatively faster than the serial algorithm. Also, the screening stage is highly data-parallel and ideal for implementation on the DAP; and it is believed that a complete DAP-based

171

substructure system would be capable of very fast response times.

To test such a system, a file of around 200,000 molecules would be required, in order that enough molecules were passed to the search stage to fill the PE array on a DAP 610 (this calculation assumes a screen-out of 98%). The existing software could be adapted easily to exploit the DAP fast-disk unit, which allows data to be read into the DAP at up to 18 Mbytes per second.

9.2.4   Other Applications of the Subgraph Isomorphism Algorithms

This book has considered the adaptation of a subgraph isomorphism algorithm for the processing of chemical structures. Although the performance levels achieved for this application were good, it is believed that higher levels of performance could be achieved for other types of graph. During the initial testing of Algorithm I, plain graphs which had no node or edge-labels were used as test data, as these were easy to generate. It was found that when the level of connectivity in the graphs was high, very high performance levels were achieved. This was because the high level of connectivity in such graphs meant that a high proportion of the run-time was spent in the relaxation procedure, which had an efficient parallel implementation.

An interesting exercise would be to study the relative performance of the DAP, compared with a serial processor, for the processing of plain graphs with varying levels of connectivity. Such graphs are used extensively in fields such as image processing (in which the DAP is used already).

Another area that could be investigated is the processing of 3-D chemical structures. The graphs representing such structures are fully connected, each edge label is the distance between the associated nodes; this distance is stored as a real number. The high connectivity of such graphs would render them ideal for processing by Algorithm I. However, this was not considered earlier because of the time-complexity involved in processing the edge-labels. The comparison of edge-labels requires real arithmetic which is not performed as efficiently on the DAP as Boolean or integer work. However, a DAP which had the co-processor hardware attached could perform such processing efficiently, and it is believed that the DAP would provide a cost-effective tool for the substructure searching of databases of 3-D molecules.

Bibliography

[1] *Cray-1 Computer System, Reference Manual Publication 2270004*. Mendota Heights, Minnesota, 1976.

[2] *DAP Series Technical Overview*. Active Memory Technology Ltd., Reading, England, October 1989.

[3] *IMS t414 Transputer Reference Manual*. Bristol, 1985.

[4] International Union of Pure And Applied Chemistry. Commission on the Nomenclature of Organic Chemistry. *Nomenclature of Organic Chemistry*, Pergamon Press, Oxford, 1979. Sections A, B, C, D, E, F and H.

[5] *The Oxford Dictionary of Current English*. Oxford University Press, Oxford, 1985.

[6] *Parallel Data Transforms*. Active Memory Technology Ltd., Reading, England, December 1988.

[7] R.A. Abagyan and V.N. Maiorov. A simple qualitative representation of polypeptide chain folds: comparison of protein tertiary structures. *Journal of Biomolecular Structure and Dynamics*, 5:1267–1279, 1988.

[8] G.W. Adamson and D. Bawden. A method of structure-activity correlation using Wiswesser Line Notation. *Journal of Chemical Information and Computer Sciences*, 15:215–220, 1975.

[9] G.W. Adamson, J.M. Bird, G. Palmer, and W. Warr. In-house chemical databases at ICI. *Journal of Molecular Graphics*, 4:165–169, 1986.

[10] G.W. Adamson, J.M. Bird, G. Palmer, and W. Warr. Use of MACCS within ICI. *Journal of Chemical Information and Computer Sciences*, 25:90–92, 1985.

[11] G.W. Adamson, J. Cowell, M.F. Lynch, A.H.W. McLure, W.G. Town, and M. Yapp. Strategic considerations in the design of a screening system for substructure searches of chemical structure files. *Journal of Chemical Documentation*, 13:153–157, 1973.

[12] M.A. Arbib, A.J. Kfourg, and R.N. Moll. *A Basis for Theoretical Computer Science*. Springer-Verlag, New York, 1984.

[13] P.J. Artymiuk, D.W. Rice, E.M. Mitchell, and P. Willett. Structural resemblance between the families of bacterial signal-transduction proteins and of G proteins revealed by graph theoretical techniques. *Protein Engineering*, 4:39–42, 1990.

[14] V. Arvind and V. Kathail. A multiple processor dataflow machine that supports generalised procedures. *Computer Architecture News*, 9:291–302, 1981.

[15] J. Ash, P. Chubb, S. Ward, S. Welford, and P. Willett. *Communication, Storage and Retrieval of Chemical Information*. Ellis Horwood, Chichester, 1985.

[16] J.E. Ash and E. Hyde, editors. *Chemical Information Systems*. Ellis Horwood, Chichester, 1975.

[17] J.H. Austin Jr. *The Burroughs Scientific Processor*, pages 1–31. Volume 2, Infotech International Limited, Maidenhead, 1979.

[18] P.A. Baker, G. Palmer, and P.W.L. Nichols. The Wiswesser Line-Formula Notation. In J. Ash and E. Hyde, editors, *Chemical Information Systems*, chapter 9, Ellis Horwood, Chichester, 1975.

[19] M.Z. Balent and J.M. Emberger. A unique chemical fragmentation system for indexing patent literature. *Journal of Chemical Information and Computer Sciences*, 15:100–104, 1975.

[20] S. Barcza, L.A. Kelly, S.S. Wahrman, and R.E. Kirschenbaum. Structured biological data in the Molecular Access System. *Journal of Chemical Information and Computer Sciences*, 25:55–59, 1985.

[21] S. Barcza, H.W. Mah, M.H. Myers, and S.S. Wahrman. Integrated chemical-biological-spectroscopy-inventory preclinical database. *Journal of Chemical Information and Computer Sciences*, 26:198–204, 1986.

[22] J.M. Barnard, editor. *Computer Handling of Generic Chemical Structures*. Gower, 1984.

[23] H.G. Barrow, A.P. Ambler, and R.M. Burstall. Some techniques for recognising structures in pictures. In S. Watanabe, editor, *Frontiers of Pattern Recognition*, pages 1–30, Academic Press, New York, 1972.

[24] D. Bawden, J.T. Catlow, T.K. Devon, J.M. Dalton, M.F. Lynch, and P. Willett. Evaluation and implementation of topological codes for online compound search and registration. *Journal of Chemical Information and Computer Sciences*, 21:83, 1981.

[25] D. Bawden and T.K. Devon. Ringdoc: the database of pharmaceutical literature. *Database*, 3(3):29–39, 1980.

[26] F.C. Bernstein, T.F. Koetzel, G.J.B. Williams, E.F. Meyer, M.D. Brice, J.R. Rogers, O. Kennard, T. Shimanouchi, and M. Tasumi. The Protein Data Bank: a computer-based archival file for macromolecular structures. *Journal of Molecular Biology*, 112:535–542, 1977.

[27] D.W. Blevins, E.W. Davis, R.A. Heaton, and J.H. Reif. Blitzen: a highly integrated massively parallel machine. *Journal of Parallel and Distributed Computing*, 8:150–160, 1990.

[28] A.T. Brint and P. Willett. Pharmacophoric pattern matching in files of 3D chemical structures. *Journal of Molecular Graphics*, 5:49–56, 1987.

[29] R.S. Cahn and O.C. Dermer. *Introduction to Chemical Nomenclature*. Butterworths, London, 1979. 5th Edition.

[30] N. Carriero and D. Gelernter. How to write parallel programs: a guide to the perplexed. *ACM Computing Surveys*, 21:323–357, 1989.

[31] B.F. Chambers, D.A. Duce, and G.P. Jones, editors. *Distributed Computing*. Academic Press, London, 1984.

[32] B.K. Chen, C. Horvath, and J.R. Bertino. Multivariate analysis and quantative structure-activity relationships. Inhibition of dihydrofolate reductase and thymidylate synthetase by quinazolines. *Journal of Medicinal Chemistry*, 22:483–491, 1979.

[33] W.F. Clocksin and C.S. Mellish. *Programming in PROLOG*. Springer-Verlag, London, 1987.

[34] S.A. Cook. The complexity of theorem-proving procedures. In *Proceedings of the 3rd Annual ACM Symposium on the Theory of Computing*, pages 151–158, Association for Computing Machinery, New York, 1971.

[35] J.E. Crowe, P. Leggate, B.N. Rossiter, and J.F.B. Rowland. The searching of Wiswesser Line Notations by means of a character-matching serial search. *Journal of Chemical Documentation*, 13:85–92, 1973.

[36] P. Cruickshank. How a computer program helped build a chemical registry. *Industrial Chemical News*, 5:36–37, 1984.

[37] J. Darlington and M. Reeve. Alice-a multiprocessor reduction machine for the parallel evaluation of functional programming languages. In *Functional Programming Languages and Computer Architectures*, pages 65–75, ACM, New Hampshire, 1981.

177

[38] C.H. Davis and J.E. Rush. *Information Retrieval and Documentation in Chemistry*. Greenwood Press, 1974.

[39] J.B. Dennis. Dataflow supercomputers. *IEEE Computer*, 13(11):48–56, 1980.

[40] R. Diamand. Applications of computer graphics in molecular biology. *Computer Graphics Forum*, 3:3–11, 1984.

[41] G.M. Downs, M.F. Lynch, G.A. Manson, P. Willett, and G.A. Wilson. Transputer implementations of chemical substructure searching algorithms. *Tetrahedron Computer Methodology*, 1:207–217, 1988.

[42] D.L. Eager, J. Zahorjan, and E.D. Lazowska. Speedup versus efficiency in parallel systems. *IEEE Transactions on Computers*, 38:408–423, 1989.

[43] M. Elder. The conversion from Wiswesser Line Notation to CIS Connection Tables. In *Proceedings of the CNA(UK) Seminar on Interconversion of Structural Representations*, Loughborough, March 1982.

[44] V. Faber, O.M. Lubeck, and A.B. White Jr. Comments on the paper 'Parallel efficiency can be greater than unity'. *Parallel Computing*, 4:209–210, 1987.

[45] H. Falk. Reaching for the gigaflop. *IEEE Spectrum*, 13(10):65–70, 1976.

[46] N.A. Farmer, J. Amos, W. Farel, J. Fehribach, and C. Zeidner. The evolution of the CAS parallel structure searching architecture. In W.E. Warr, editor, *Chemical Structures: The International Language of Chemistry*, pages 283–296, Springer,Heidelberg, 1988.

[47] N.A. Farmer and M.P. O'Hara. A new source of substance information from Chemical Abstracts Service. *Database*, 3(4):10–25, 1980.

[48] G. Fayolle, P.J.B. King, and I. Mitrani. On the execution of programs by many processors. In *Proceedings of the 9th International Symposium on Computational Performance and Measurement Evaluation*, pages 217–228, 1983.

[49] E.A. Feigenbaum and P. McCormick. *The Fifth Generation - Artificial Intelligence and Japan's Challenge to the World.* Michael Joseph, London, 1983.

[50] A. Feldman and L. Hodes. An efficient design for chemical structure searching, 1. The screens. *Journal of Chemical Information and Computer Sciences*, 15:147–152, 1975.

[51] W. Fisanick. The Chemical Abstracts Service generic chemical (Markush) structures storage and retrieval capability. I. Basic concepts. *Journal of Chemical Information and Computer Sciences*, 30:145–154, 1990.

[52] P.M. Flanders. A unified approach to a class of data movements on an array processor. *IEEE Transactions on Computers*, C-31:809–819, 1982.

[53] P.M. Flanders and D. Parkinson. Data mapping and routing for highly parallel processor arrays. *Future Computing Systems*, 2:183–224, 1987.

[54] M.J. Flynn. Some computer organisations and their effectiveness. *IEEE Transactions on Computers*, C-21:948–960, 1985.

[55] E.C. Freuder. Structural isomorphism of picture graphs. In C.H. Chen, editor, *Pattern Recognition and Artificial Intelligence*, pages 248–256, Academic Press, New York, 1976.

[56] R. Fugmann. *The IDC System*, pages 195–226. Ellis Horwood, Chichester, 1975.

[57] D.D. Gajski and J.K. Peir. Essential issues in multiprocessor systems. *Computer*, 18(6), 1985.

[58] M.R. Garey and D.S. Johnson. *Computers and Intractability, A Guide to the Theory of NP Completeness.* Freeman, 1979.

[59] L. Goebels, A.J. Lawson, and J.L. Wisniewski. AUTONUM: System for computer translation of structural diagrams into IUPAC-compatible names. 2. Nomenclature of chains and rings. *Journal of Chemical Information and Computer Sciences*, 31:216–224, 1991.

179

[60] R.L. Graham. Bounds on multiprocessing anomalies and packing problems. In *Proceedings AFIPS Spring Joint Computer Conference*, pages 205–217, 1972.

[61] J.L. Gresling. *Mathematical Structures for Computer Science.* Freeman, 1987.

[62] H.M. Grindley. *3-D Distance Comparisons in Protein Secondary Structure Motifs for Substructure Searching Refinement.* Master's thesis, University of Sheffield, 1988.

[63] J.L. Gustafson, G.R. Montry, and R.E. Benner. Development of parallel methods for a 1024-processor hypercube. *Journal of Scientific Statistical Computation*, 9:609–638, 1988.

[64] T.R. Hagadone. Computer graphics based definition of generic chemical structures. In J.M. Barnard, editor, *Computer Handling of Generic Chemical Structures*, pages 159–161, Gower, 1984.

[65] T.R. Hagadone. Current approaches and new direction in the management of in-house chemical structure databases. In W.E. Warr, editor, *Chemical Structures: The International Language of Chemistry*, pages 22–43, Springer, Heidleberg, 1988.

[66] C. Hansch and M. Yoshumoto. Structure-activity relationships in immunochemistry. 2. Inhibition of complement by benzamidines. *Journal of Medicinal Chemistry*, 17:1160–1167, 1974.

[67] I. Hayes, editor. *Specification Case Studies.* Prentice Hall, London, 1987.

[68] P. Heidelberger and K.S. Trivedi. Queuing network models for parallel processing with asynchronous tasks. *IEEE Transactions on Computers*, C-31:1099–1109, 1982.

[69] D. Heller. A survey of parallel algorithms in numerical linear algebra. *SIAM Review*, 20:740–777, 1978.

[70] A.J. Hey. Reconfigurable transputer networks, practical concurrent computation. *Philosophical Transactions of the Royal Society of London*, 326:395–410, 1988.

[71] W.D. Hillis. *The Connection Machine*. M.I.T., Cambridge, Massachusetts, 1985.

[72] W.D. Hillis. The Connection Machine. *Scientific American*, 256(6):86–93, 1987.

[73] R.W. Hockney and C.R. Jessope. *Parallel Computers 2, Architecture, Programming and Algorithms*. Adam Hilger, Bristol, 1988.

[74] L. Hodes. Selection of descriptors according to discriminacy and redundancy-application to chemical structure searching. *Journal of Chemical Information and Computer Sciences*, 16:88–93, 1976.

[75] D.K. Isenor and S.G. Zaky. Fingerprint identification using graph matching. *Pattern Matching*, 19:113–122, 1986.

[76] C. Jenson. Taking another approach to supercomputing. *Datamation*, 24(2):159–175, 1978.

[77] C. Jochum, J. Gasteiger, and I. Ugi. The principle of minimum chemical distance (PMCD). *Angewandte Chemie*, 19:495–505, 1980.

[78] P. Jochum and T. Worbs. A multiprocessor architecture for substructure search. In W. Warr, editor, *Chemical Structures: The International Language of Chemistry*, pages 279–282, Springer, Heidelberg, 1988.

[79] T.M. Johns. Chemical graphics: bringing chemists into the picture. In *Graphics for Chemical Structures: Integration with Text and Data*, pages 18–28, 1987.

[80] T.A. Jones. Interactive computer graphics: FRODO. *Methods in Enzymology*, 115:157–171, 1985.

[81] J. Kao, V. Day, and L. Watt. Experience in developing and building an in-house molecular information and modelling system. *Journal of Chemical Information and Computer Sciences*, 25:129–135, 1985.

[82] L. Kohn and Margulis. The i860 64-bit supercomputing microprocessor. In *Supercomputing '89*, pages 450–456, ACM, Reno, USA, 1989.

[83] Y. Kudo and H. Chihara. Chemical retrieval system for searching generic representations. 1. A prototype system for the gazetted list of existing chemical substances of Japan. *Journal of Chemical Information and Computer Sciences*, 23:109–117, 1983.

[84] S.A. Kuschel and A.C. Page. Augmented relaxation labelling and dynamic relaxation labelling. *IEEE Transactions on Pattern Analysis and Machine Intelligence*, 4:676–682, 1982.

[85] U. Lang. Simple CPU-benchmarks in Fortran 77 on different machines. *Supercomputer*, 6:26–31, 1988.

[86] A.P. Lurie. CAS and MDL registry systems at Eastman Kodak company. In W.E. Warr, editor, *Chemical Structures: The International Language of Chemistry*, pages 77–78, Springer, Heidelberg, 1988.

[87] M.F. Lynch. Screening large chemical files. In J. E. Ash and E. Hyde, editors, *Chemical Information Systems*, pages 177–194, Ellis Horwood, Chichester, 1975.

[88] M.F. Lynch, J.M. Barnard, and S.M. Welford. Computer storage and retrieval of generic chemical structures in patents. Part 1. Introduction and general strategy. *Journal of Chemical Information and Computer Sciences*, 21:148–150, 1981.

[89] M.F. Lynch and P. Willett. The automatic detection of chemical reaction sites. *Journal of Chemical Information and Computer Sciences*, 18:154–159, 1978.

[90] Y.C. Martin, M.G. Bures, and P. Willett. Searching databases of three-dimensional structures. In K. B. Lipkowitz and D. B. Boyd, editors, *Reviews in Computational Chemistry*, pages 213–263, VCH, New York, 1990.

[91] D. May. The influence of VLSI technology on computer architecture. *Philosophical Transactions of the Royal Society of London*, 326:377–393, 1988.

[92] T.K. Ming and S.J. Tauber. Chemical structure and substructure search by set reduction. *Journal of Chemical Documentation*, 11:47–51, 1970.

[93] M. Minsky and S. Papert. On some associative, parallel and analog computations. In *Associative Information Technologies*, pages 609–638, Elsevier, New York, 1971.

[94] E.M. Mitchell. *Protein Secondary and Tertiary Structure Searching in Files of 3-D Atomic Co-ordinates taken from the Protein Data Bank.* PhD thesis, University of Sheffield, 1988.

[95] E.M. Mitchell, P.J. Artymiuk, D.W. Rice, and P. Willett. Use of techniques derived from graph theory to compare secondary structure motifs in proteins. *Journal of Molecular Biology*, 212:151–166, 1990.

[96] T.E. Moock, B. Christie, and D. Henry. MACCS-3D - a new database system for three-dimensional molecular models. In D. Bawden and E. M. Mitchell, editors, *Chemical Information Systems: beyond the structure diagram*, pages 42–49, Ellis Horwood, Chichester, 1990.

[97] C.N. Mooers. Zatocoding applied to mechanical organisation of knowledge. *Journal of American Documentation*, 2:20–32, 1951.

[98] H.L. Morgan. The generation of a unique description for chemical structures - a technique developed at Chemical Abstracts Services. *Journal of Chemical Documentation*, 5:107–113, 1965.

[99] T. Nakayama and Y. Fujiwara. Computer representation of generic chemical structures by an extended block-cutpoint tree. *Journal of Chemical Information and Computer Sciences*, 23:80–87, 1983.

[100] J. Page, R. Theisen, and K. Kuhl. The Walter Reed Army Institute of Research chemical information system. *ACS Symposium Series*, 84, 1978.

[101] D. Parkinson. Performance analysis in a 4096 processor environment. *Journal of Systems and Software*, 1(2):11–15, 1986.

[102] D. Parkinson and H. M. Liddell. The measurement of performance on a highly parallel system. *IEEE Transactions on Computers*, 32:32–37, 1983.

[103] R.H. Perrot. *Parallel Programming*. Addison-Wesley, Workingham, England, 1987.

[104] S.L. Peyton-Jones. *The Implementation of Functional Languages*. Prentice-Hall, Englewood Cliffs, New Jersey, 1987.

[105] D.J. Polton. Installation and operational experiences with MACCS (molecular access program). *Online Review*, 6:235–242, 1982.

[106] J.L. Potter, editor. *The Massively Parallel Processor*. MIT Press, Cambridge, MA, 1985.

[107] M.J. Quinn. *Designing Efficient Algorithms for Parallel Computers*. McGraw-Hill, New York, 1987.

[108] B. Randell, editor. *The Origins of Digital Computers - Selected Papers*. Springer-Verlag, Berlin, 1975.

[109] R.C. Read and D.G. Corneil. The graph isomorphism disease. *Journal of Graph Theory*, 1:339–363, 1977.

[110] S.F. Reddaway. Achieving high performance applications on the DAP. In C.R. Jesshope and K.D. Reinartz, editors, *CONPAR 88*, Cambridge University Press, Cambridge, 1989.

[111] S.F. Reddaway. DAP - A Distributed Array Processor. In *1st Annual Symposium on Computer Architecture*, IEEE, ACM, New York, 1972.

[112] A.J. Richard and L.B. Kier. Structure-activity analysis of hydrazide monoamine oxidase inhibitors using molecular connectivity. *Journal of Pharmaceutical Sciences*, 69:124–126, 1980.

[113] D.J. Rosenkrantz, R.E. Stearns, and P.M. Lewis. An analysis of several heuristics for the travelling salesman problem. *SIAM Journal of Computuation*, 6:563–581, 1977.

[114] S. Rossler and A. Kolb. The GREMAS system - an integral part of the IDC system for chemical documentation. *Journal of Chemical Documentation*, 10:128–134, 1970.

[115] A.H. Sameh. Numerical parallel algorithms - A survey. In D.J. Kuck, D.H. Lawrie, and A.H. Sameh, editors, *High Speed Computer and Algorithm Organisation*, pages 207–228, Academic Press, New York, 1977.

[116] J.T. Schwartz. Ultra-computers. *ACM Transactions on Language Systems*, 2:484–521, 1980.

[117] C.L. Seitz. The Cosmic Cube. *Communications of the Association of Computing Machinery*, 28:22–23, 1985.

[118] J.A. Sharp. *An Introduction to Distributed and Parallel Computing*. Blackwell, London, 1987.

[119] K. Shenton, P. Norton, and E.A. Fearns. Generic searching of patent information. In W. Warr, editor, *Chemical Structures: The International Language of Chemistry*, pages 169–178, Springer, Heidelberg, 1988.

[120] E.H. Sussenguth. A graph theoretic algorithm for matching chemical structures. *Journal of Chemical Documentation*, 5, 1963.

[121] G.S. Taylor. Arithmetic on the ELXSI System 6400. In *6th Annual Symposium on Computer Architecture*, pages 110–115, IEEE, New York, 1983.

[122] W.G. Town. Microcomputers and information systems. *Chemistry in Britain*, 25(11):1118–1120, 1989.

[123] A. Tucker. *Applied Combinatorics*. Wiley, New York, 1980.

[124] J.R. Ullmann. An algorithm for subgraph isomorphism. *Journal of the Association for Computing Machinery*, 23:31–42, 1976.

[125] B. van't Riet, L.B. Kier, and H.L. Elford. Structure-activity relationships of benzohydroxamic acid inhibitors of ribonucleotide reductase. *Journal of Pharmaceutical Science*, 69:856–857, 1980.

[126] A. von Scholley. A relaxation algorithm for generic chemical structure screening. *Journal of Chemical Information and Computer Sciences*, 24:235–241, 1984.

[127] S.B. Walker. Development of CAOCI and its use in ICI Plant Protection Division. *Journal of Chemical Information and Computer Sciences*, 23:3–5, 1983.

[128] D.L. Waltz. Applications of the Connection Machine. *Computer*, 20(1):85–97, 1987.

[129] D. Weininger. SMILES, a chemical language information system. 1. Introduction to methodology and encoding rules. *Journal of Chemical Information and Computer Sciences*, 28:31–36, 1987.

[130] R. Wilensky. *LISPcraft*. Norton, London, 1984.

[131] P. Willett. A review of chemical structure retrieval systems. *Journal Of Chemometrics*, 1:139–155, 1987.

[132] P. Willett. *Similarity and Clustering in Chemical Information Systems*. Research Studies Press, Letchworth, 1987.

[133] P. Willett and V. Winterman. A comparison of some measures of intermolecular structural similarity. *Quantative Structure-Activity Relationships*, 5:18–25, 1986.

[134] R.J. Wilson. *Introduction To Graph Theory*. Oliver and Boyd, Edinburgh, 1972.

[135] W.T. Wipke. Three dimensional substructure search. In *186th National Meeting of the American Chemical Society*, Washington D.C., August 1983.

[136] W.T. Wipke, S. Krishnan, and G.I. Ouchi. Hash functions for rapid storage and retrieval of chemical structures. *Journal of Chemical Information and Computer Sciences*, 18:32–37, 1977.

[137] W.T. Wipke and D. Rogers. Rapid subgraph search using parallelism. *Journal of Chemical Information and Computer Sciences*, 24:255–262, 1984.

[138] J.L. Wisniewski. AUTONUM: System for computer translation of structural diagrams into IUPAC-compatible names. 1. General design. *Journal of Chemical Information and Computer Sciences*, 30:324–332, 1990.

[139] W.J. Wiswesser. *A Line-Formula Chemical Notation*. New York, 1954.

Appendix A

Iterative Ullmann Code

There follows a description of the Ullmann algorithm as reported in Ullmann's original paper [124]. There are several mistakes in this, these have been underlined (Mitchell [94] noted that a problem did exist with the original algorithm, as coded by Brint [28]; however the errors, and the modifications necessary to correct these errors, have never been published).

All variables are the same as those used in the thesis; two further temporary variables are used:

- F is a P_β bit binary vector which records which columns of M have been used at an intermediate state of the computation: $F_i = 1$ if the ith column has been used.

- H is an integer vector of P_α elements which records which columns of M have been used at which depth: $H_d = k$ if the kth column has been selected at depth d.

Step 1. $M := M^0$; $d := 1$; $H_1 := 0$;
        for all $i := 1, \ldots, P_\alpha$ set $F_i := 0$; {should be $i = 1, \ldots, P_\beta$}
        refine M; if exit FAIL then terminate algorithm;

Step 2. If there is no value of j such that $m_{dj} = 1$ and $F_j = 0$ then go to step 7;

$M_d := M$;

if $d = 1$ then $k := H_1$ else $k := 0$;

Step 3. $k := k + 1$;

if $m_{dk} = 0$ or $F_k = 1$ then go to step 3;

for all $j \neq k$ set $m_{dj} := 0$;

refine M; if exit FAIL then go to step 5;

Step 4. If $d < P_\alpha$ then go to step 6 else give output to indicate that an isomorphism has been found;

Step 5. If there is no $j > k$ such that $m_{dj} = 1$ and $F_j = 0$ then go to step 7;

$M := M_d$; {this line should be executed before the previous line}

go to step 3;

Step 6. $H_d := k$; $F_k := 1$; $d := d + 1$;

go to step 2;

Step 7. If $d = 1$ then terminate algorithm;

$F_k := 0$; $d := d - 1$; $M := M_d$; $k := H_d$;

{the previous line should read: $F_k := 0$; $d := d - 1$; $k := H_d$; $F_k := 0$;}

go to step 5;

189

Appendix B

Query Structure Diagrams

There follows a list of the structure diagrams corresponding to the query structures used in the substructure searching experiments reported in Chapter 6.

Query numbers 8, 11, 34, 37, 43, 45 and 57 contained more than 16 atoms and could not be processed by Algorithm II because of insufficient array store.

Structure 1

Structure 2

Structure 3

Structure 4

Structure 5

Structure 6

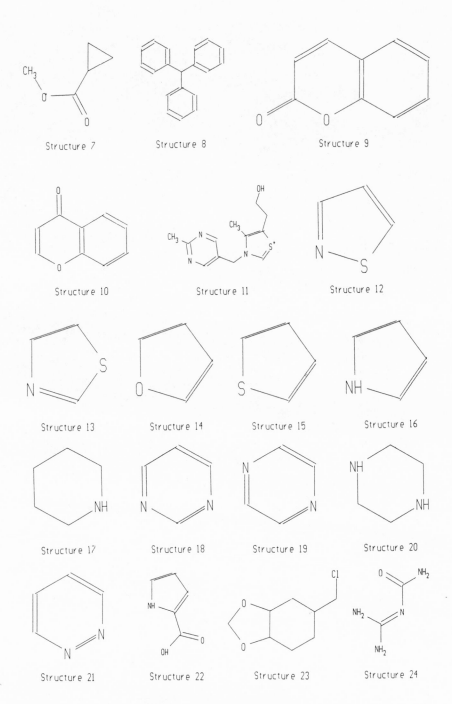

Structure 7

Structure 8

Structure 9

Structure 10

Structure 11

Structure 12

Structure 13

Structure 14

Structure 15

Structure 16

Structure 17

Structure 18

Structure 19

Structure 20

Structure 21

Structure 22

Structure 23

Structure 24

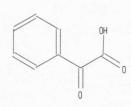

Structure 25

Structure 26

Structure 27

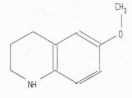

Structure 28

Structure 29

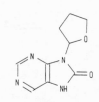

Structure 30

Structure 31

Structure 32

Structure 33

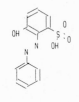

Structure 34

Structure 35

Structure 36

Structure 37

Structure 38

Structure 39

192

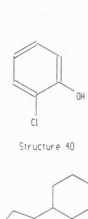

Structure 40

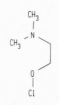

Structure 41

Structure 42

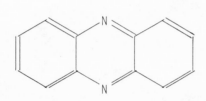

Structure 43

Structure 44

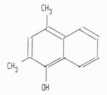

Structure 45

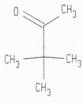

Structure 46

Structure 47

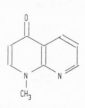

Structure 48

Structure 49

Structure 50

Structure 51

Structure 52

Structure 53

Structure 54

Structure 55

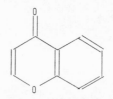

Structure 56

Structure 57

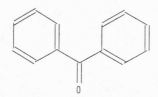

Structure 58

Structure 59

Structure 60

Structure 61

Structure 62

Structure 63

Structure 64

Appendix C

Abbreviations

ALU	Arithmetic Logic Unit
AMT	Active Memory Technology Ltd.
APAL	Array of Processors Assembly Language
CAS	Chemical Abstracts Service
CPU	Central Processing Unit
DAP	Distributed Array Processor
ES	Edge Size
JANET	Joint Academic NETwork
MCS	Maximal Common Subgraph
MCU	Master Control Unit
MIMD	Multiple Instruction stream, Multiple Data stream
MIPS	Million Instructions Per Second
NP	Non-deterministic Polynomial time
P	deterministic Polynomial time
PDT	Parallel Data Transform
PE	Processing Element
RISC	Reduced Instruction Set Computer
SIMD	Single Instruction stream, Multiple Data stream
SSE	Secondary Structure Element
VLSI	Very Large Scale Integration
WLN	Wiswesser Line-formula Notation

Index

Adjacency matrix 37
Algorithmic parallelism 100
Amdahl's law 71
Array store 78, 79
ASN 65
Autonum 13

Bit-string 132
Blitzen 59

CAS online 31
Chemical abstracts service 8, 9, 21
Complexity 40, 105, 153
Connection machine 54, 55
Connection table 14, 36, 131, 132

DAP 54, 58, 73, 127, 152
Dataflow 48, 53, 61
Drip-feed algorithm 120

Flynn's taxonomy 57
Fortran Plus 73, 80, 82, 133, 135
Fragmentation codes 6

Generic structure 19
GENSAL 20
Geometric parallelism 100
Granularity 67
Graph theory 35, 172

Hybrid algorithm 119

Information system 5

Konigsburg walk 35

Linear notation 7
Long vector mode 88

Loosely/tightly coupled processors 55

M matrix 94
MACCS 33
Macromolecule ranking 3, 144
Majority voting 110
Markush DARC 21
MARPAT 21
Masking 87, 109
Maximal common subgraph 18
MCU 74, 108
MIMD processing 59, 110
Morgan algorithm 22

Nomenclature 6, 12
NP-completeness 2, 33, 43, 101

Parallel processing 2, 33, 46, 49, 74, 100
PDT 91, 108, 121, 134, 136, 139
Pipelining 51
POSSUM 145
Processing element 75
Protein 144

Relaxation 30, 93, 101, 111
RISC processor 77

Screening search 23, 171
Set reduction 29, 93
Shore's taxonomy 64
SIMD processing 55, 58, 100
Similarity search 17
SMILES 7
Speedup 69
Structure diagram 6, 10, 38, 190
Structure search 16, 21, 39
Subgraph isomorphism 2, 40
Substructure search 2, 17, 23, 39, 127

Transputer 56, 60
Tree compression 112
Tree pruning 103, 115
Tree search 101
Turing machine 42

Ullmann algorithm 93, 110, 129, 145, 188

Von Neumann 1, 46

Wiswesser line notation 7, 13

Z 63